Multilevel Environmental Governance

Multilevel Environmental Governance

Managing Water and Climate Change in Europe and North America

Edited by

Inger Weibust

Assistant Professor, Norman Paterson School of International Affairs, Carleton University, Canada

James Meadowcroft

Professor, School of Public Policy and Administration, Carleton University, Canada

Edward Elgar

Cheltenham, UK • Northampton, MA, USA

Published by
Edward Elgar Publishing Limited
The Lypiatts
15 Lansdown Road
Cheltenham
Glos GL50 2JA
UK

Edward Elgar Publishing, Inc.
William Pratt House
9 Dewey Court
Northampton
Massachusetts 01060
USA

A catalogue record for this book
is available from the British Library

Library of Congress Control Number: 2013949799

This book is available electronically in the ElgarOnline.com Social and Political Science Subject Collection, E-ISBN 978 0 85793 925 8

MIX
Paper from responsible sources
FSC www.fsc.org FSC® C013056

ISBN 978 0 85793 924 1 (cased)

Typeset by Columns Design XML Ltd, Reading
Printed and bound in Great Britain by T.J. International Ltd, Padstow

Contents

Figure and tables vii
Contributors viii
Abbreviations x
Acknowledgements xii

PART I THEORETICAL PERSPECTIVES ON MULTILEVEL GOVERNANCE

1. Introduction 3
Inger Weibust
2. Too many levels or just about right? Multilevel governance and environmental performance 15
Daniel J. Fiorino

PART II MULTILEVEL GOVERNANCE OF WATER RESOURCES

3. Subsidiarity as a 'scaling device' in environmental governance: the case of the European Union 39
David Benson and Andrew Jordan
4. Multilevel governance and the politics of environmental water recoveries 58
B. Timothy Heinmiller
5. Playing a zero sum game: sharing water between jurisdictions in federations 80
Inger Weibust

PART III MULTILEVEL GOVERNANCE OF CLIMATE CHANGE MITIGATION

6. Climate governance in the European Union multilevel system: the role of cities 111
Kristine Kern

7. Bottom-up versus top-down: the evolving American climate policy odyssey 131
Barry G. Rabe
8. Institutional strength, intergovernmental relations, and national climate policy coordination: Australia and Canada compared 154
David Gordon and Douglas Macdonald
9. Allocating greenhouse gas emission reductions amongst sectors and jurisdictions in federated systems: the European Union, Germany and Canada 181
Douglas Macdonald

PART IV FINDINGS ON EFFECTIVENESS AND GOVERNANCE PATTERNS

10. Ensuring the effectiveness of European Union environmental law: from supranational law making to multilevel enforcement 215
Marc Pallemaerts
11. What is multilevel environmental governance? When does it work? 249
Inger Weibust

Index 271

Figures and tables

FIGURE

6.1. Cities in EU climate governance 114

TABLES

2.1. An illustrative matching of environmental problems to governance levels 28
7.1. Intergovernmental involvement in American climate change policy, 1975–2013 135
8.1. Elements contributing to institutional strength/weakness of IGR 158
8.2. Projected impact on jurisdictional GDP relative to business as usual (BAU) scenario in 2020 160

Contributors

David Benson
Department of Politics, University of Exeter, Cornwall Campus
Penryn, Cornwall (UK)

Daniel J. Fiorino
Department of Public Administration and Policy, American University
Washington DC (US)

David Gordon
Department of Political Science, University of Toronto
Toronto, Ontario (Canada)

B. Timothy Heinmiller
Department of Political Science, Brock University
St Catharines, Ontario (Canada)

Andrew Jordan
School of Environmental Sciences, University of East Anglia
Norwich (UK)

Kristine Kern
Department of Economic and Social Sciences, University of Potsdam
Potsdam (Germany)

Douglas Macdonald
School of the Environment, University of Toronto
Toronto, Ontario (Canada)

James Meadowcroft
School of Public Policy and Administration, Carleton University
Ottawa, Ontario (Canada)

Marc Pallemaerts
Institute for European Environmental Policy
Brussels (Belgium)

Barry G. Rabe
Gerald R. Ford School of Public Policy, University of Michigan
Ann Arbor, Michigan (US)

Inger Weibust
Norman Paterson School of International Affairs, Carleton University
Ottawa, Ontario (Canada)

Abbreviations

3G	third generation
ACESA	American Clean Energy and Security Act
ACF	Apalachicola-Chattahoochee-Flint river system
BAU	business as usual
CCME	Canadian Council of Ministers of the Environment
CCS	carbon capture and storage
CPRS	Pollution Reduction Scheme
CAP	Common Agricultural Policy
CAF	Council of the Australian Federation
COAG	Council of Australian Governments
CEM	Council of Energy Ministers
COF	Council of the Federation
ETS	Emissions Trading Scheme
EAP	Environmental Action Programme
EKC	Environmental Kuznets Curve
EPA	Environmental Protection Agency
ECSC	European Coal and Steel Community
EC	European Commission
EEC	European Economic Community
EEA	European Environment Agency
EU	European Union
FITFIR	first in time, first in right
FMM	First Ministers meeting
GL	gigalitre
GHG	greenhouse gases
IMPEL	European Network for the Implementation of Environmental Law
IGA	intergovernmental agreement
IGR	intergovernmental relations
IWRM	integrated water resource management
IJC	International Joint Commission
JMM	joint Ministers meetings
MRBB	Mackenzie River Basin Board
MGGRA	Midwestern Greenhouse Gas Reduction Accord

NAICC	National Air Issues Coordinating Committee
NAICC-CC	National Air Issues Coordinating Committee–Climate Change
NAISC	National Air Issues Steering Committee
NCCP	National Climate Change Process
NCCS	National Climate Change Secretariat
NETS	National Emissions Trading Scheme
NETT	National Emissions Trading Taskforce
NCSL	National Governors Association and National Conference of State Legislatures
NGO	non-governmental organization
PPWB	Prairie Provinces Water Board
RGGI	Regional Greenhouse Gas Initiative
REA	Renewable Energy Act
RET	renewable energy target
RPS	renewable portfolio standards
RBMP	River Basin Management Plans
SEA	Single European Act
SSR	South Saskatchewan River
TEU	The Treaty of European Union
ToA	The Treaty on Amsterdam
UNFCCC	United Nations Framework Convention on Climate Change
UK	United Kingdom
US	United States
VCR	Voluntary Challenge and Registry
WCI	Western Climate Initiative
WFD	Water Framework Directive
WWBA	Western Water Board Agreement
WGCCW	Working Group on Climate Change and Water

Acknowledgements

This volume is a publication of the Centre for European Studies (CES) at Carleton University in Ottawa, Canada. The Centre is grateful for financial support from Carleton University and from the European Union, the latter based on a grant to CES, which was designated an EU Centre of Excellence by the European Commission in 2006.

We also acknowledge support from the Social Sciences and Humanities Research Council (SSHRC) of Canada for funding the research that formed the basis of some of the chapters in this volume.

We are particularly grateful to our copy editor, Jane Freeland. We thank all of the authors for the collective effort that went into producing a coherent approach and for their attention to our suggestions, requests and inquiries.

Of course, this publication reflects the views of the authors for each chapter; none of the sponsoring organizations is responsible for the content, and the contents of this publication can in no way be taken to reflect the views of the European Union.

Inger Weibust and **James Meadowcroft**

Disclaimer

PART I

Theoretical perspectives on multilevel governance

1. Introduction

Inger Weibust

Climate change is the most pressing environmental problem facing the world today; it represents the mother of all global collective action problems. Freshwater management, while not requiring a global solution, represents a ubiquitous problem. Worldwide, problems of water supply and quality are expected to be exacerbated by global climate change. A range of tools is available to address each of these problems yet they remain poorly managed, particularly in the case of climate change mitigation. The stumbling block is not an absence of technical solutions but, rather, political and institutional factors. Thus it matters how successful different forms of governance have been in tackling these problems.

Policy studies, political science and public administration have all, in recent years, witnessed a turn towards governance, away from government, as the relevant object of study. Government implies, 'hierarchical decision-making structures and the centrality of public actors … the former denotes the participation of public and private actors, as well non-hierarchical forms of decision-making' (Kohler-Koch and Rittberger 2006, 28). There is a sense that policy making is now too complex and too messy to be contained within the vessels of elected governments and bureaucracies:

> No single actor, public or private, has all [the] knowledge and information required to solve complex, dynamic and diversified problems; no actor has sufficient overview to make the application of particular instruments effective; no single actor has sufficient action potential to dominate unilaterally in a particular governing model (Kooiman 1993, 4).

National governments, in particular, have lost authority. From above their sovereign powers are circumscribed by international organizations, or supranational ones such as the European Union (EU). From below they have either devolved tasks to lower-level governments or have fewer resources to control activities at the subnational level.

This book takes as its focus patterns of environmental policy making in systems of multilevel governance (MLG). It examines environmental policies on climate and water in several such systems and critically examines widely held assumptions about them. By taking a comparative perspective, these chapters seek to shed light on which aspects of MLG seem to be generalizable across systems and which are *sui generis*. The European Union in particular, has been the focus of much theorizing, which assumes that the EU is unique (Kohler-Koch and Rittberger 2006, 33). However, any claims to uniqueness will have greater validity if made in comparison to similar objects.

In the case of environmental policy making, is it generally true that the national state in MLG has become weaker or is losing power to other levels? Are government authority and hierarchy becoming less important to environmental policy in federal-type systems? Lastly what can we conclude about variants of MLG and effects on environmental policy making?

DEFINITIONS OF MULTILEVEL GOVERNANCE

As the term 'multilevel governance' has become more commonplace, the ambiguity of its meaning has increased. This book will refer to two distinct meanings of the term. First, multilevel governance is used as a descriptive term for governance in federal systems that goes beyond the usual two levels that define federal systems. Second, it refers to the particular conception of multilevel governance embodied in the theory of multilevel governance originated by Liesbet Hooghe (1996) and Gary Marks (1993) to describe the evolving governance of the European Union (Hooghe and Marks 2001). In Marks' definition (1993, 392), multilevel governance is 'a system of continuous negotiation among nested governments at several organizational tiers.'

Although the term 'multilevel governance' originated as a theory about the development of the European Union, for environmental policy, it is now employed simply as a descriptive term, often entirely removed from the European context.[1] 'Multilevel governance' has filled the need for a term to describe the many levels of government involved a wide variety of policy areas. The terms 'federal system' or 'federation' typically focus on the activities of and relationship between the federal level of government and subnational units, such as provincial or state governments. (In the case of the European Union the two corresponding levels are the institutions of the European Union and the Member State governments.) This typology, however, is inadequate to capture what happens in policy areas where local governments, as well as federal and sub-national levels, are important. Furthermore, a focus on federal structures does not incorporate the international level of governance, which is key in many policy areas. Climate change is a paradigm case of a policy where all four levels play a part in policy making and where the interaction between levels can affect activity at any level (Rabe 2007). This definition of multilevel governance can be thought of as 'federalism plus:' the two levels that define federal systems, augmented by the international/supranational level and/or the local level.

HOOGHE AND MARKS' THEORY OF MULTILEVEL GOVERNANCE

Hooghe and Marks' theory of multilevel governance clearly captures the idea of governance involving multiple levels, beyond federal and sub-national. However it goes considerably beyond that simple definition and beyond its vertical emphasis. Their theory entails broader empirical claims: predictions about the evolution of governance, as well as normative claims for multilevel governance.

Hooghe and Marks' starting point was the unitary state. Studying European Union Member States, they observed that the nation state was weakening as power shifted upwards to the European Union but also downwards to regional governments. Further they argued that, not only did other levels matter more in governance, but that the sense of a hierarchical relationship between these levels was dissolving:

> While national arenas remain important arenas for the formation of national government preferences, the multi-level governance model rejects the view that subnational actors are nested exclusively within them. Instead, sub-national actors operate in both national and supranational arenas … National governments … share, rather than monopolize, control over many activities that take place in their respective territories. (Hooghe and Marks 2001, 4)

The assumption that higher levels of government were placed in a hierarchical order over the lower, having authority over them, was also weakening.

In addition to a weakening of vertical, top-down authority, Hooghe and Marks also saw a weakening of government authority in general, vis-à-vis other actors in society, hence their preference for the term 'governance' over 'government.' Government control is being replaced by steering. Thus, in their conception of multilevel governance, vertical authority between levels of government is weakened but so is the horizontal authority of all levels of government over other societal actors. Thus for Hooghe and Marks all levels of government are becoming less important in policy making as non-state actors become more important.

In addition to claiming that this weakening of authority is a trend, they go further and make the normative claim that these developments represent *progress*. Hooghe and Marks (2003, 233) state that:

> [c]entralized authority – command and control – has few advocates. Modern governance is – and, according to many, should be – dispersed across multiple centers of authority. But how should multi-level governance be organized?

This is familiar territory to economists studying fiscal federalism and optimal jurisdiction, who ask 'which level of government is best suited to which tasks? Why?' (Dafflon 2006). Based on the theory of optimal jurisdiction, environmental governance may not be well suited to fragmentation and decentralization (Dalmazzone 2006). Many environmental problems involve externalities or spillovers: governments may cause environmental damage outside their own boundaries, which economists would predict governments don't have much incentive to address.

Hooghe and Marks' Multilevel Governance has been criticized for several reasons (George 2004). Jordan (2001, 201) presents some of the common criticisms. First, he characterizes Hooghe and Marks' concept of MLG as 'thick description' rather than theory, because it does not offer a causal account for why these changes are occurring. He goes on to claim that MLG theory significantly overstates the autonomy of subnational actors in the EU. In particular, he cautions about drawing conclusions about subnational governments' actual *influence* on policy based on their activity or degree of mobilization in Brussels.

CLAIMS ABOUT DECENTRALIZATION

Multilevel governance theory also makes normative claims about the decentring and fragmentation of governing and government authority: 'dispersion of governance across multiple jurisdictions is both more efficient than and normatively superior to the central state monopoly' (Hooghe and Marks 2004, 16). At present decentralization is in fashion among policy makers and academics (Treisman 2007, 1). For example, since 1987 the subsidiarity principle has, in theory, determined the division of competencies in European Union environmental policy making: that decisions should be made at the lowest possible level of government.[2] The general claims typically advanced for political decentralization include: greater responsiveness and accountability to the public; more policy innovation; and better adaptation to local conditions (Treisman 2007).

In the US, the individual states have long been hailed as 'laboratories of democracy', a term coined by US Supreme Court Justice Louis Brandeis. Brandeis postulated a process by which state governments were engaged in innovation and experimentation, to be followed by the subsequent diffusion of best policies to other states (Tarr 2001). Treisman (2007, 234), examining social policy, goes further, arguing that in federal systems, such as Canada, social policy innovation has occurred at the subnational level because subnational governments have been able to block federal policy innovations.

Scholarship sceptical of decentralization, particularly with respect to the environment and natural resource management, has emerged in response. Treisman's (2007) wide-ranging critique of political decentralization argues that there is little basis for the claims made in favour of decentralization and that empirical studies are inconclusive. With regard to innovation and environmental policy, Lowry (1992) and Rabe (1998), both present findings that contradict the claim that decentralization is most conducive to environmental policy innovation.

Examining state level environmental policy in the United States (US), Lowry (1992) found that states were more likely to innovate and implement more stringent standards in those areas of pollution control where the US federal government set minimum national standards. Absent federal standards, state governments had complete autonomy to innovate, but hesitated to do so, because of competitiveness concerns. Comparing American state governments and Canadian provincial governments for innovation in pollution control, Rabe (1998) found that, despite their greater autonomy, provincial governments engaged in very little innovation. In contrast American state governments, which are far more constrained by federal oversight, were much more likely to innovate in pollution control.

American research on natural resource management at the state or local level has found that maximizing autonomy at lower levels results in relatively worse results in environmental protection. Comparing the management of state and federal forests in the US, Koontz (2002) found that federal forests were managed with a view to environmental protection, while state forests prioritized timber yields.

In a similar vein, studies of local natural resource management and public participation have sounded a cautionary note. Layzer (2008) undertook a systematic comparison of four cases of collaborative local ecosystem-based management with three cases using more conventional regulatory approaches. She found that none of the cases using collaborative approaches was 'likely to conserve or restore the landscapes they aim to protect' (Layzer 2008, 6). Furthermore, 'initiatives in which goals emerged out of conventional politics have yielded greater-than-expected environmental benefits … [They] mitigated the disparity in power between development and environmental interests' (Layzer 2008, 5). In collaborative arrangements, the need to achieve consensus led to the avoidance of controversy and avoidance of imposing short run costs on interests favouring development. Examining several cases of public participation in local environmental decision making, Beierle and Cayford (2002, 62) concluded that the regulatory context was an important determinant of success: 'public participation – even when done well – is not a substitute for the regulatory power, political will and money required to get things done.'

THE RETURN OF HIERARCHY IN GOVERNANCE

More recently the role of government authority and hierarchy in governance arrangements has received renewed scrutiny. Examining new environmental policy instruments in nine European countries, Jordan et al (2005) found few cases of true 'governance,' without any government authority. Most new policy instruments relied upon at least some state involvement (Jordan et al. 2005, 477). Government authority and hierarchy may provide benefits which negotiation-based network governance does not. Effective policy in many spheres may depend upon *the shadow of hierarchy*. The shadow of hierarchy can be the background presence of government authority in arrangements for industry self-regulation that are motivated by the credible threat of government regulation (Héritier and Eckert 2008, 117). These are examples of policy and governance where government appears to play a limited role. However, in these cases the presence, albeit limited, of government authority appears to be a key determinant of efficacy. Thus there is a difference in kind, not just degree, between governance with little direct government involvement and governance entirely lacking in government authority or hierarchy.

Hooghe and Marks (2003) are correct in their assertion that command and control by central authorities has few advocates at present. However they may be too quick to jettison government authority and hierarchy as a whole. A more nuanced representation of hierarchy is possible: negotiation versus hierarchy may represent a false dichotomy. Fritz Scharpf (1997) surveyed a variety of possible governance arrangements for policy making, including the hierarchical direction of decisions and negotiation in networks. Scharpf (1997, 197) made the distinction between 'hierarchical authority structure and the *actual use* of hierarchical direction, in order to override the decision preferences of other actors … Within a hierarchical authority structure, therefore, it is indeed possible that actual interactions will have the character of negotiations or of unilateral action.' Thus, in a hierarchical structure, actual decisions may be made by negotiation, most of the time. The possibility of hierarchical decision making (imposition) remains an option, kept in reserve. However, Scharpf noted that the reverse is not true: a hierarchy free network system of governance cannot generate majority voting or hierarchical decision making.

THE EMPIRICAL MATERIAL: CLIMATE POLICY AND WATER POLICY

To broaden the basis of comparison, the book compares multilevel environmental governance in several federal systems. Our principal cases are the European Union, Canada and the US. For some policies we have additional case material from Germany and Australia. Our ability to make generalizations about these systems is strengthened by focusing the discussion on two key areas of environmental policy: climate policy and water policy.

Based on first principles, we would assume that governments would treat climate issues differently from water issues. The globe shares a single climate: thus solving climate change necessitates cooperation. Much of the international relations literature assumes that the problems of coordination and cooperation increase with the number of governments involved. Thus if we proceed from an economist's rational actor model, we would not expect much climate action at the subnational and local levels of government. The costs of coordinating such a large number of governments are high and opportunities for free riding or cheating are substantial. Furthermore, the question of 'leakage' is very important in adopting climate change policies: policies to reduce emissions in one place will be futile if economic activity (and attendant emissions) just shift elsewhere. Thus, achieving reductions through local level or state level policy is predicted to be more difficult that achieving reductions through national policy.

While water scarcity is a ubiquitous problem, encountered in many places around the world, it is not a global problem in the same way that climate is. Spain may learn from California about managing scarce water resources. However, sustainably managing Spain's water resources does not require California's active participation or cooperation. Local, regional or national solutions to water policy are possible in a way they are not with climate change.

What the two policy areas have in common is a confrontation with the limits of natural systems, and the need to impose real constraints on human activity as a result. Mitigating climate change will require real reductions in greenhouse gas (GHG) emissions. Similarly, managing water resources often requires real cuts in water consumption, or tough decisions over reallocating water between users. A common strategy for avoiding hard choices in politics in general is to make the pie bigger. For example, conflict over distribution of wealth can be diffused through economic growth that makes everyone's share bigger, even if relative shares don't change. Growing the total pie is not a long-term option in climate or water policy however. Global GHG emissions need to be capped, then reduced. Similarly, in many cases, the water supply 'pie' is shrinking in relative terms, because of population growth and reduced supplies as a result of climate change. Thus, while the geographic scale of climate change mitigation and water management are different, the need to confront a hard natural limit is the same.

Some of the claims of MLG theory can be tested. If we take a snapshot today, what most accurately describes environmental governance in federal systems such as the EU and US? Are they top-down, bottom-up or something else entirely? Examining these systems, do we observe, over time, that hierarchies are weakening, that authority is fragmenting? Where we observe variation in governance structure across federal systems, what can we learn about structures and policy effectiveness? These are the sorts of questions which the essays in this volume attempt to address.

The contribution by Daniel J. Fiorino surveys comparative research on relationships between institutional structures and environmental outcomes. He argues that there is no simple correspondence between particular institutional arrangements and positive environmental outcomes. His chapter also reflects on federal/state interactions around environmental governance in the United States over the last forty years to draw some conclusions about the significance of multilevel governance for policy outcomes.

Taking up the theme of the optimal governance scale for environmental policy, David Benson and Andrew Jordan examine how the European Union sought to address the question of assigning jurisdictional responsibility in water quality policy. After 1987, the European Union turned to the principle of subsidiarity to resolve these conflicts. The chapter explores how the subsidiarity principle came to be applied and its consequences for the trajectory of the EU's environmental policy.

Turning to water supply, Timothy Heinmiller's chapter compares irrigation policy reforms at the subnational level in Canada, Australia and the US. All systems have seen a shift away from promotion and expansion of irrigation towards more conservation-focused water policies. Given the federal nature of these countries, the transition has taken place within multilevel institutions, which have significantly shaped the design and extent of sustainability reforms. Alberta, New South Wales and California are compared to illustrate how the different institutions in each federation have facilitated or impeded the general transition from expansionary to sustainability approaches in managing water for irrigation.

Inger Weibust's chapter compares the institutional arrangements for allocating the waters of transboundary rivers in the United States, Canada and the European Union. These regimes vary significantly in terms of the autonomy that lower level governments have in allocating water and their recourse to authoritative dispute resolution. They also show variation on policy outcomes, such as guaranteeing minimum stream flows for ecological purposes.

Turning to climate change, Kristine Kern's chapter examines the role that cities are playing in the multilevel governance of climate within the European Union. Kern describes how cities are implementing and shaping European climate policy. Kern identifies three distinct types of climate policy Europeanization at the urban level: 'hierarchical Europeanization,' 'cooperative Europeanization,' and 'horizontal Europeanization,' and considers their significance for climate policy.

Barry G. Rabe's chapter explores the ways in which complex interactions between different levels of government have shaped the US response to climate change. He suggests a fourfold typology of state/federal interactions, arguing that over time there has been movement from a 'symbolic policy 1975–1997' (low state and federal involvement), to 'state domination 1998–2007' (with high state and low federal involvement) and then to 'contested federalism 2008–2013'. And he asks whether the coming decade may see a shift towards the final category of 'federal domination' (with high federal and low state involvement).

David Gordon and Douglas Macdonald seek to explain the puzzle of why Australia was successful in crafting a national climate change policy, while Canada has not been. The two countries have comparable greenhouse gas emissions profile, economic structure and political institutions. Both are relatively decentralized federal systems. Gordon and Macdonald attribute the difference in outcome at least in part to differences in the institutions of intergovernmental relations within each of the two countries.

Douglas Macdonald examines efforts by Germany, the European Union, and Canada to allocate cuts in greenhouse gas emissions. A basic challenge facing climate change mitigation policy makers is deciding which sources will reduce greenhouse gas emissions by what portion of the total reduction target. In a unitary state, policy makers do this by negotiating with sources. At the global level, although the same need to allocate amongst sources exists, the focus is upon allocation amongst states. In federated systems, however, particularly when sources are geographically concentrated and important to local economies, there is necessarily a simultaneous focus upon both allocation amongst sources and amongst the subnational governments in the federation. Two federated systems, Germany and the EU, have achieved success in reaching agreement on allocation of total GHG reduction cost amongst sources and jurisdictions. Canada has not. Macdonald explains the varying success of these efforts with reference to institutions, the hierarchical relationships among them, and the motivations of the governments involved in each case.

Marc J. Pallemaerts discusses the mechanisms at the European Union's disposal for enforcing EU environmental law. Since the early 1970s, the EU has developed an impressive body of environmental legislation that prevails over the national laws of its Member States. Yet it is well known that the effectiveness and enforcement of this legislation continue to pose major problems in all Member States as well as, paradoxically, at the level of the EU institutions themselves. The European Community Treaty provides for both centralized and decentralized enforcement mechanisms, with a key role for the European Commission, the European Court of Justice, and the national courts of the Member States. Citizens and civil society organizations, too, have a role to play in enforcing EU environmental law, a role which was actively encouraged by the Commission during the 1980s and 1990s, but seems to be more circumscribed recently.

The concluding chapter returns to the overarching questions posed at the beginning of this introduction, and discusses the chapters in light of those questions.

NOTES

1. See for example, (Selin and VanDeveer 2009) and (Winter 2006).
2. However one of the leading scholars of European environmental law considered that this principle had never been applied in decisions about what to regulate at the European Union level (Krämer, 1998, 73).

REFERENCES

Beierle, Thomas C. and Jerry Cayford (2002), *Democracy in Practice: Public Participation in Environmental Decisions*, Washington DC: Resources for the Future.

Dafflon, Bernard (2006), 'The assignment of functions to decentralized government: from theory to practice,' in Ehtisham Ahmad and Giorgio Brosio (eds), *Handbook of Fiscal Federalism*, Cheltenham, UK; Northampton, MA, USA: Edward Elgar, pp. 27–305.

Dalmazzone, Silvana (2006), 'Decentralization and the environment,' in Ehtisham Ahmad and Giorgio Brosio (eds), *Handbook of Fiscal Federalism*, Cheltenham, UK; Northampton, MA, USA: Edward Elgar, pp. 459–477.

George, Stephen (2004), 'Multi-level Governance and the European Union,' in Ian Bache and Matthew Flinders (eds), *Multi-level Governance*, Oxford: Oxford University Press, pp. 107–126.

Héritier, Adrienne and Sandra Eckert (2008), 'New modes of governance in the shadow of hierarchy: Self-regulation by industry in Europe,' *Journal of Public Policy* **28**(1), 113–138.

Hooghe, Liesbet (1996), *Cohesion Policy and European Integration: Building Multi-Level Governance*, Oxford: Oxford University Press.

Hooghe, Liesbet and Gary Marks (2001), *Multi-Level Governance and European Integration*, Lanham: Rowman & Littlefield.

Hooghe, Liesbet and Gary Marks (2003), 'Unraveling the central state, but how? Types of multi-level governance,' *American Political Science Review* **97**(2), 233–243.

Hooghe, Liesbet and Gary Marks (2004), 'Contrasting visions of multi-level governance,' in Ian Bache and Matthew Flinders (eds) *Multi-level Governance*, Oxford: Oxford University Press, pp. 15–30.

Jordan, Andrew (2001), 'The European Union: an evolving system of multi-level governance … or government?' *Policy & Politics* **29**(2), 193–208.

Jordan, Andrew, Rüdiger K. W. Wurzel and Anthony Zito (2005), 'The rise of 'new' policy instruments in comparative perspective: Has governance eclipsed government?' *Political Studies* **53**(3), 477–496.

Kohler-Koch, Beate and Berthold Rittberger (2006), 'Review article: The "governance turn" in EU studies,' *Journal of Common Market Studies* **44**(1), 27–49.

Kooiman, Jan (1993), *Modern Governance: Government-Society Interactions*, London: Sage.

Koontz, Tomas M. (2002), *Federalism in the Forest: National versus State Natural Resource Policy*, Washington DC: Georgetown University Press.

Krämer, Ludwig (1998), *EC Treaty and Environmental Law*, 3rd ed., London: Sweet & Maxwell.

Layzer, Judith A. (2008), *Natural Experiments: Ecosystem-Based Management and the Environment*, Cambridge: MIT Press.

Lowry, William R. (1992), *The Dimensions of Federalism: State Governments and Pollution Control Policies*, Durham, London: Duke University Press.

Marks, Gary (1993), 'Structural policy and multilevel governance in the EC,' in A. Cafruny and G. Rosenthal (eds), *The State of the European Community*, Boulder: Lynne Rienner, pp. 391–411.

Rabe, Barry G. (1998), 'Federalism and entrepreneurship: Explaining American and Canadian innovation in pollution prevention and regulatory integration,' *Policy Studies Journal* **27**(2), 288–306.

Rabe, Barry G. (2007), 'Beyond Kyoto: Climate change policy in multilevel governance systems,' *Governance* **20**(3), 423–444.

Scharpf, Fritz (1997), *Games Real Actors Play: Actor-Centered Institutionalism in Policy Research*, Boulder: Westview Press.

Selin, Henrik and Stacy D. VanDeveer (eds) (2009), *Changing Climates in North American Politics: Institutions, Policymaking and Multilevel Governance*, Cambridge: MIT Press.

Tarr, G. Alan (2001), 'Laboratories of democracy? Brandeis, federalism, and scientific management', *Publius* **31**(1), 37–46.

Treisman, Daniel (2007), *The Architecture of Government: Rethinking Political Decentralization*, Cambridge: Cambridge University Press.

Winter, Gerd (ed.) (2006), *Multilevel Governance of Global Environmental Change*, Cambridge: Cambridge University Press.

2. Too many levels or just about right? Multilevel governance and environmental performance

Daniel J. Fiorino

Martin Jänicke and colleagues define environmental policy capacity as 'a society's ability to identify and solve environmental problems' (Jänicke 1997, 1; see also Jänicke and Weidner 1997; Jänicke 1992; Weidner and Jänicke 2002; Weidner 2002). This comparative research on many countries stresses the importance of economic, social and political characteristics as they affect this policy capacity. Part of the argument in this research is that the policies or instruments relied upon are less important in determining capacity than are the 'structural preconditions' within which problem solving occurs. One such set of structural preconditions relates to the distribution of authority at different levels of government and their contributions to effective problem solving.

To gain a picture of the role of multilevel governance, it is necessary to examine the range of factors that may be associated with effective problem solving at the national level. Among these are: the level of economic development in a society, usually expressed in terms of per capita income; the regime type, as defined by degrees of democracy in a system; and the institutional characteristics of a system, including the relationships among political institutions (such as parliamentary compared to presidential or separation of powers systems, electoral rules, the judicial system, and the authority and role of subnational levels of government). The purposes of this chapter are (1) to consider what the research on national environmental performance suggests about the effect of multilevel governance (with performance as a surrogate for capacity) and (2) to propose issues and general directions for future research on multilevel governance.

The premise of this chapter and indeed of much of this volume, is that patterns in multilevel governance may have important implications for environmental performance and problem solving capacity. This is a critical question in the current policy environment. Many national systems of governance, such as the United States (US), Canada, Germany, and Australia, stake their success on the effective operation and integration of strong subnational units of government. Moreover, the demands of modern governance are increasing the role of regional and global institutions, which share many characteristics with established national systems based on federalism. The most obvious such case is the European Union (EU), which retains national authority but with a strong overlay of supranational governance. In the case of the EU, the lessons of multilevel governance in more established federal systems may offer guidance on how to design and manage policies and institutions. Even beyond the EU, the need for regional and global cooperation makes a study of multilevel governance compelling and relevant.

The first part of the chapter examines research on national environmental performance. The dependent variable in most of this research is some measure of either outputs or outcomes that reflect policy capacity. Although there are differences between the concepts of performance and capacity, they are both used somewhat interchangeably for the purposes of this discussion. Later the chapter considers two specific questions in light of the research on performance:

(1) Do patterns of multilevel governance play out differently at various stages of the policy cycle?
(2) Are some governance levels better at responding to specific problems than others?

For both questions, the intent is to set out directions for research rather than provide definitive answers.

EXPLAINING NATIONAL ENVIRONMENTAL PERFORMANCE

Why do some nations perform better on most environmental indicators than others? To what extent do different patterns in multilevel governance affect environmental performance? Can these effects be isolated from the other explanations for performance, such as levels of economic development, relative degrees of democracy or other institutional characteristics? This section considers research on the effects of variations in subnational governance (i.e., the effects of federal compared to more unitary systems) and its relevance to performance. It offers no clear answers, because the effects of patterns of multilevel governance are difficult to isolate from other explanations for environmental performance. However, reviewing this research does help in explaining how variations in governance patterns may influence national environmental performance in a variety of situations. It also sheds light on the challenges of attempting to disentangle the effects of multilevel governance on environmental performance.

This chapter draws extensively upon a review of the literature on national performance conducted by the author (Fiorino 2011). Before moving to the categories of explanations, it is worth noting the central features of this research (Fiorino 2011, 369–71). The dependent variable is usually defined as some version of environmental performance. Some studies focus on policy outputs, such as laws enacted, institutions created, commitments made, or policies adopted in some form. Other studies rely on measures of policy outcomes, such as air and water pollution, forests and other protected habitat, greenhouse gas emissions or the health effects of contaminants. Within the outcomes category some studies use trends over time as a dependent variable; others use environmental conditions at a point in time (an example is Emerson et al. 2010). These variations complicate the task of drawing broad conclusions, so this chapter gives some attention to definitions of the dependent variables and the implications for explanations of performance.

Four kinds of explanation may be identified in the research on national performance. One examines the relationship between economic growth, usually defined in terms of per capita income and performance on a range of output and outcome indicators. Much of this research focuses on the notion of an Environmental Kuznets Curve (EKC), in which many forms of air and water pollution increase in the early stages of growth but begin to decline absolutely later (examples are Dasgupta et al. 2001; Dasgupta et al. 2005; Grossman and Krueger 1995; Stern 2004; Levinson 2002; Cole et al. 1997; Dinda 2004). The asserted basis for this inverted U-shaped curve is that pollution increases in the early stages of growth, industrialization and urbanization as an almost inevitable result of increased energy use, manufacturing, mobility and consumption. However, at some point, as societies become more affluent, and their citizens become more concerned about the quality of life, governments intervene to reduce pollution levels. Political demands for action, combined with a greater ability to fund and apply technologies, lead to reductions in pollution levels and to many other forms of environmental degradation, despite economic growth and higher per capita incomes (Perrings and Ansuategi 2000).

In one sense such findings support assertions that economic growth is not incompatible with environmental protection. Since the emergence of the modern environmental movement in the late 1960s, most economically affluent societies have adopted pollution control policies and to some extent have 'decoupled' growth from environmental damage. These findings should be qualified however. First, this decoupling has occurred largely in the case of health-related air and water pollution for which the effects are salient politically and the subject of voter pressures. For other indicators, such as greenhouse gas emissions, ecosystem or species protection and resource consumption, the decoupling process and income levels at which it occurs is less clear or nonexistent. Second, nearly all of this EKC research has been conducted in the more advanced industrial economies rather than in developing economies. The relationship of growth to environment may well exhibit different patterns in these countries (an exception is Miah et al. 2011). Third, some of the research has found other patterns in the relationship of income growth to environmental performance, with some researchers finding evidence of an S-shaped curve or other pattern more than an inverted U (Stern 2004; Levinson 2002). Still, this body of research suggests that growth is not necessarily associated with environmental damage across the board. It suggests that emerging economies may 'tunnel through' the EKC with appropriate policies and institutions (Munasinghe 1999).

A second set of explanations focuses on the effects of relative degrees of democracy. In the 1970s, when the environmental critique focused on the need to control or stop economic growth in recognition of the limits of the biosphere, some critics argued that democratic systems were unequal to the task. Imposing limits on growth, affluence and consumption, it was argued, is difficult in a democracy. Research over the last few decades has not supported this assertion however. Democratic political systems have been found in most studies to have achieved higher levels of environmental protection than their more authoritarian counterparts. Early research into this area found that increasing levels of freedom in a country are associated with reductions in air pollution (Barrett and Graddy 2000; see also Frederiksson and Wollschied 2007; Gallagher and Thacker 2008; Bernauer and Koubi 2009; Payne 1995). A recent comprehensive study concluded that 'a rise in democracy reduces environmental degradation and improves environmental performance' (Li and Reuveny 2006, 936). This conclusion is based on an analysis of five outcome indicators: carbon dioxide and nitrogen dioxide emissions; organics in water; deforestation; and land degradation.

A related body of research found that democracies exhibit higher levels of commitment to international environmental problem solving, in the form of greater participation in international agreements, organizations and reporting (Neumayer 2002; Recchia 2002). Although this is based on measures of commitment (an output variable) rather than actual environmental outcomes, it does provide another perspective on the performance of democracies. Other studies have found democracies to be more effective in committing to actions on climate change but less so in fulfilling them in terms of performance (Battig and Bernauer, 2009). There is also evidence that performance varies based on institutional characteristics (Frederiksson and Wollscheid 2007) and on the *specific* environmental outputs or outcomes being studied (Duwel 2010).

The empirical research generally supports the relative superiority of democratic regimes in environmental performance, although with reservations. The theoretical case for the positive role of democracy rests on several arguments. Among these are a free flow of information, so that environmental problems may come to light; opportunities for mobilizing societal interests to demand pollution control programs; freedom to support research and technology innovation; and the ability of voters to hold elected officials accountable. It is also likely that in democratic regimes there are linkages to other aspects of economic and political development. More affluent societies for example, which also tend to be more democratic, are also likely to exhibit more stable and responsive political institutions, an effective civil service, and the rule of law (Dasgupta *et al.* 2006; Esty and Porter 2005). These overall governance capacities have been the subject of some attention in the literature, but deserve more study. Multilevel governance issues should be seen in the context of these capacities.

A third set of explanations, and the most relevant one to the discussion in this chapter, is the effects of the institutional characteristics of national political systems. Do parliamentary systems like the United Kingdom perform better in terms of their ability to identify and respond to environmental problems than those based on a separation of powers, such as the United States? Do smaller, issue oriented 'green' parties stand a better chance of exerting legislative influence in electoral systems using proportional representation than in winner-take-all political systems? Do systems characterized by a high number of veto points (i.e., the more pluralist ones) exhibit lesser levels of environmental performance than those in which institutions are better able to facilitate political consensus and policy integration (i.e., the more corporatist political systems)? More specific to my purpose – Are federalist systems any more or less capable of delivering a high level of environmental performance than unitary ones that lack strong subnational governance?

Although a fair amount of empirical research has been conducted into these questions, the challenges in drawing definitive conclusions are formidable. As would be expected for example, green parties are more likely to gain legislative seats and influence in electoral systems based on proportional representation (Frederiksson and Millimet 2004; Knill et al. 2010). This is why we see legislative 'greens' in northern Europe but not in the US Congress. The effects of this representation are less clear, although it appears that environmental issues are more likely to emerge on legislative agendas in these systems (Harrison and Sundstrom 2010).

One institutional analysis of the effects of multilevel governance on performance describes the complexity of the relationships (Wälti 2004). Are more federal (strong multilevel) systems better environmental performers than more unitary (weak multilevel) ones? Focusing on two air pollutants (nitrogen dioxide and sulfur dioxide), Wälti concludes that multilevel governance interacts with two other system characteristics – income and corporatism. Like the other studies, she finds positive relationships between performance, income and corporatism. At least for these two air pollutants, wealthier and more corporatist countries perform better. Being multilevel (strongly federalist) correlates positively with environmental outcomes, but the specific effects vary, depending on income and the extent of corporatism. In less federalist systems, income has a more significant effect then corporatism. In more federalist systems, such as the US or Canada, corporatism matters more than does economic status as measured by income.

The explanations for these findings lie in the interactions between the effects of corporatism, income, and multilevel governance. Income and strong multilevel governance both correlate positively with performance. It appears that, in wealthier nations that are strongly federalist, 'more economic and fiscal resources may be channeled to subnational levels of government' (Wälti, 2004, 606). With respect to the multilevel-corporatist relationship, previous studies concluded that all interest groups benefit when there are multiple access points. Environmental interests have more leverage in more federalist systems and economic interests in more unitary ones. It may be that the former respond more to local concerns, allow environmentalists to monitor implementation, and provide options to environmentalists when other government levels oppose environmental actions. Similarly Poloni-Staudinger (2008, 424) concludes that devolution has positive effects by enhancing responsiveness, legitimacy and enforcement.

As for the effects of parliamentary systems relative to presidential ones, the best answer is that it all depends. Conventional theory suggests that parliamentary systems are more able to form coalitions and implement policy change than presidential/separation of powers ones. The larger number of veto points in a system such as the US poses barriers to change in response to voter demands. In the 1970s however, it was the US that moved rapidly to adopt an environmental regulatory regime, with the parliamentary systems in Great Britain and Japan following behind. When more conservative leadership gained power in all three countries in the 1980s, the many veto points in the American system made attempts to roll back regulatory standards more difficult (Vogel 1993).

What has shown up in research is that neo-corporatist political systems, because of their ability to integrate across policy sectors and interests, perform better environmentally. In one of the more comprehensive studies, Scruggs (2003) found that the overall configuration of power matters more than do specific institutional factors. In a category of independent variables labelled 'socio-economic institutions,' he assesses the quality of relationships among economic, environmental, and other interests with government, and their roles in policy making. The distinction is between the pluralist systems, like the US or Canada, and neo-corporatist ones like Germany and Sweden. He finds that 'countries characterized by strong, centralized interest groups and a more 'consensual' approach to policymaking have enjoyed better environmental performance than countries where economic groups are less comprehensively organized and policymaking is less consensual (Scruggs 2003, 123). Jahn (1998, 124) concludes that 'a better environmental performance is correlated with neo-corporatist and strong Social Democratic politics.' In an air quality study, Crepaz (1995) finds lower pollution levels in corporatist than in pluralist systems. The former are more goal oriented, accommodative and able to establish structures that reduce policy uncertainty. A more recent analysis focusing on policy outputs (Liefferink et al. 2009, 92) concludes that 'collective action problems inherent in environmental policy can be solved more easily in neo-corporatist "closed shops" based on trust and long-term reciprocity.' These findings are consistent with theoretical and case analyses of environmental performance (examples are Dryzek 1997 and Lafferty and Meadowcroft 2000).

Yet a fourth area of research is relevant to studying the effects of multilevel governance. Its focus is on the diffusion of environmental policies and innovations among countries. The premise of this research is that, whatever the effects of domestic political, economic and other factors, countries (and units of government within countries) may influence and learn from each other (Lenschow et al. 2005). These studies tend to use measures of environmental outputs – policies, laws, programs – as the dependent variable, with measures of policy diffusion across national lines as explanatory ones.

In a study of the diffusion of green plans, for example, Jörgens (2004) proposes four determinants of successful diffusion: the existence of communication channels, innovation properties, problem structure and capacities for incorporating innovations. Jänicke and Jacob (2004, 41) argue that countries seeking economic advantage through innovation will provide examples for others. In what Vogel (1997) terms the 'California effect,' one jurisdiction with economic clout can set stringent standards that lead others to adopt them. Empirical support for this effect may be found in two recent studies, both of which relied on output measures as the dependent variable. Consistent with much of the previous research, one found that per capita income, neo-corporatism and environmental pressures (as measured by industrial carbon dioxide emissions per capita and population density) are associated positively with performance. Even more important was membership in the European Union (EU) (Liefferink et al. 2009). Similarly another study (also using output measures) concluded that EU membership 'significantly increases' the number of environmental policies adopted (Holzinger et al. 2011, 328; also see Börzel 2000).

This discussion suggests several conclusions about the effects of multilevel governance on environmental performance. One is that governance levels should be seen in the context of larger economic, institutional and cultural characteristics. Based on the limited amount of research that has been done, it appears that the degree to which a system is unitary or federal does not independently influence policy outcomes. It is but one of many factors that influences policy in combination with these other factors. To evaluate the effects of multilevel governance, one would also need to understand its interactions with the vertical distribution of powers, electoral rules and representation in legislatures, the role of courts in refereeing relationships among levels of government, and (as Scruggs 2003, terms it) the overall configuration of power in a system. More broadly such factors as income levels, economic structure, state political culture and, perhaps, more specific governance characteristics (such as legislative professionalism or rules of legislative organization and procedure at national and subnational levels) could matter.

Two sets of findings from this research are especially important in assessing the effects of multilevel governance. One is the extent to which more or less federalist systems interact with or are embedded in overall configurations of power, as analysed in the neo-corporatist research. The second is the effects of leadership, lesson-drawing and standard-setting in a multilevel system. Do the positive effects of EU membership play out within federalist systems as well? Do leaders bring up laggards in a federal system, especially one that is as tightly coupled as the US?

DIRECTIONS FOR FUTURE RESEARCH ON MULTILEVEL GOVERNANCE

As with many institutional characteristics, the effects of multilevel governance are difficult to isolate from other explanations for environmental performance. The effects will vary depending on the issue, the state of public opinion, the quality of leadership and many other factors. To explore the question of the effects of multilevel governance further, this part of the chapter considers two questions. First, can we develop general propositions about the effects of multilevel governance based on the stage in the policy process being studied? Second, are problems more suited to being addressed and solved at one governance level than at others? These questions expand the discussion by considering the ways in which problem solving is situation dependent. In this first case, it is the dependence on, or fit with, a stage of policy making. In the second, it is the fit between the problems and the policy capacities that exist at different levels of governance.

Stages in the Policy Cycle

Models of public policy typically specify a series of stages at which policy making occurs. A well-known environmental policy text describes six such stages in the policy cycle: (1) agenda setting; (2) formulation; (3) legitimation; (4) implementation; (5) evaluation; and (6) change (Vig and Kraft 2010). This six-stage model takes us from the point at which issues are first seen as emerging on policy agendas through to the later stages of policy adoption and revision. Different interests, information and institutions play roles at the various stages in the policy cycle. The stage of policy formulation involves an array of policy experts, issue networks, interest groups and government institutions in the process of developing alternative policy tools and strategies. Legislatures are likely to be the dominant institutions in the process of policy legitimation (although courts and executives may be important as well). Regulatory or other public agencies at various governance levels are likely to play a key role in the implementation stage. Policy evaluation and change will involve many institutions, actors and sources of data and analysis.

Of course as with any social science model, this policy cycle is but a simplified version of a far more complex reality. One stage blends into the others and the boundaries between them are often unclear. Moreover policy making rarely occurs in a linear manner: it is more likely to resemble a dynamic systems model with feedback loops and other complications. Nevertheless, the policy cycle serves the purpose of providing a framework for thinking about the effects of multilevel governance as environmental policies are developed, enacted, evaluated and revised. The discussion above suggests that it helps to think about multilevel governance specifically, at various stages in the policy cycle, rather than in terms of the effects on policy across the board.

The effects of multilevel governance will vary according to the stage in the policy cycle being studied. Four stages in the policy cycle are especially important and worth distinguishing: (1) the ways in which interests gain access and shape policy agendas; (2) the ability to effect change in policies, especially when they involve diffuse benefits and concentrated costs, as typically is the case for environmental issues; (3) the capacity for actually implementing policy change; and (4) the ability to learn from experience and revise policy in response to what is learned. These of course, are the steps of agenda setting, formulation, implementation, and evaluation/revision in the policy cycle. As a result it makes sense to make more of an effort to link the study of the effects of multilevel governance with what is known about policy processes.

Based on the earlier discussion, here are some general statements about the effects of patterns of multilevel governance as it relates to these stages in the policy process. These may be used to develop more specific research questions and policy proposals on multilevel governance.

Access and agenda setting. Stronger multilevel systems give more access to policy makers for a range of interests, especially when combined with proportional representation and a vertical separation of powers. Opportunities for bottom-up policy change are better in multi-level than in more unitary systems. Winner-take-all electoral rules will offset this effect to some degree, by allowing smaller, single-issue-party access.

Formulation. The effects will likely be mixed. Because there is another set of veto points added to the process of changing policy, strong multilevel systems may find major transitions (e.g., from fossil fuels to renewables or from subsidized to price rationed water use) to be difficult, much as conventional thinking held it to be for parliamentary relative to separation of power systems. On the other hand, increased access and the greater number of access points at which policy change may be initiated and influenced can offset this effect. Just as the characteristics of the national system in the US made it difficult to enact climate change legislation in 2009, federalism made it possible for California to adopt its own trading system in the Global Warming Solutions Act of 2006.

Implementation. Stronger multilevel governance will make national implementation more difficult. Once policies are adopted, they work their way through a complex system in which authority is delegated to subnational levels. National implementation mechanisms must also be coordinated or integrated with regional/state/provincial ones (for example, consider the need for the US to coordinate a national cap-and-trade program with the Regional Greenhouse Gas Initiative or a state program such as California's). On the other hand, delegating policy implementation below the national level may allow for more responsive and effective application of policies, because local and regional preferences will play a stronger role. The question then becomes one of the relative effects of having stronger national laws and institutions (as in the US) or having relatively stronger subnational institutions (as in Canada or Australia).

Evaluation-Revision. Strong multilevel systems are likely to exhibit and reinforce the 'lock-in effect' that occurs in check-and-balance systems like the US. Once adopted, policies are difficult to change (Vogel 1993). From an environmental perspective, this was an advantage for the US during the Ronald Reagan years, when the administration tried to roll back regulatory standards. With respect to policy learning, federalism reinforces the rigidity of separation of powers and makes it hard to modernize environmental policy. Horizontal sharing of power adds to the inflexibility and rigidity of regulatory policies in particular (Fiorino 2001). The effects of governance patterns on innovation are difficult to predict, although one study concludes that the closer interdependence among national and state environmental agencies in the US is more conducive to policy innovation than in Canadian federalism, where the provinces have more autonomy (Rabe 1999).

Matching Governance Levels to Problems

In addition to distinguishing among stages in the policy cycle, it is necessary to consider the characteristics of the problem at hand in evaluating the effects of multilevel governance. Environmental problems differ from many other kinds of policy issues in their close connection to a physical space. Among the problems having such a connection, for example, are damages to local or regional ecosystems, a contaminated aquifer, or a threatened species. Other issues affect a broader geographic area and need to be addressed at higher levels of governance. A major reason for enacting strong national pollution control legislation in the US during the 1970s was that threats to air and water quality extended beyond the authority of individual states. By contrast, climate change is a global issue. Lower levels of governance surely play a role in reducing emissions and an even greater one in strategies for adaptation, but it is difficult to imagine any long-term solutions to climate change without national and global engagement.

This section of the chapter offers two issues to consider about the relationships between governance levels and problems. The first task is to specify, as a starting point, seven levels of environmental governance. Four – local, state-provincial, national and global – represent the standard governance levels at which most policy issues are managed. Increasingly, however, there is the recognition that these are not suited to many kinds of environmental problem. As a result, regional governance institutions and policies have emerged. One such set of institutions has been devised to fill gaps between local and state-provincial governance. A second responds to a need for governance capacity between state-provincial and national levels and a third has emerged to address problems that transcend national capacity but are not fully global. For the purposes of this discussion, these three will be termed the micro, meso, and macro levels.

Consider the three regional governance levels and examples of each. Local ecosystem planning and watershed organizations are an example of micro-regional governance. Also captured in the concept of 'civic environmentalism,' they describe efforts to compensate for the limitations of top-down, technology-based regulation by engaging local stakeholders in a more diverse, bottom-up form of problem solving (John 2004). These also include older, established agencies such as the South Coast Air Quality District in California and South Florida Water Management District. Meso-governance (interstate but subnational) includes the Regional Greenhouse Gas Trading Initiative, Chesapeake Bay Commission and Northeast States for Coordinated Air Use Management. International macro-regional governance would include, as examples, the EU, the Convention on Long-Range Transboundary Air Pollution and the Great Lakes Regional Collaboration (for a discussion of one such institution, see Rabe and Gaden 2009).

Regional institutions differ from the standard governance levels in several respects. The most obvious is that they are often established in response to a specific problem or challenge. Their creation is driven by needs that are perceived as not being met or beyond the capacity of existing governance institutions. A second difference is that they are typically an improvization in which funding, leadership, institutional support, legitimacy and relationships with the public are based on the situation. A third difference is that they are even more dependent on the standard levels of governance than those other levels typically are on each other. Regional governance capacities are created by existing institutions in response to recognized gaps in policy capacity. They will only be as effective as the institutions that create them – and provide them with resources, authority and legitimacy – want them to be. An area for future research is to assess the strengths and limitations of these as well as other levels of governance.

Table 2.1 carries this line of thought a bit further by matching the seven governance levels with a list of eight typical environmental problems. The check marks in each cell are the author's admittedly subjective assessment of the levels of governance that are substantially involved in responding to and managing each problem. The Chesapeake Bay for example, is a multi-state resource. The causes of Bay pollution are diverse and complex, directly involving and affecting five states, the District of Columbia and several local jurisdictions (National Academy of Public Administration 2007). Many remaining causes of Bay pollution (urban storm water, air deposition, septic systems and agricultural run-off) must be addressed by local governments through zoning, infrastructure, land use planning and other tools. Because it affects a large region and so many jurisdictions, state and national governments have created meso-regional institutions like the Chesapeake Bay Commission.

The other problems illustrate different levels of engagement at various governance levels. In the case of managing the possible environmental and health effects of nano-technology, some local governments (e.g. Cambridge, Massachusetts and Berkeley, California) have adopted or considered rules on managing and reporting on materials, but there is little recognized need for involvement below the national and global levels. Regulation of chemical use and safety has largely been the focus of national and macro-regional institutions (such as the European Union), but American states have become more involved as the US, nationally, is seen as lagging on many chemicals issues and a complex and multi-faceted problem such as climate change requires the involvement of all governance levels. The point is not so much the assessments offered in the table but, rather, the need to think systematically about relationships among levels of governance and environmental problems.

Multilevel Governance: Solution or Complication?

A theme of this chapter is that it is not possible to draw general conclusions about the effects of different patterns of multilevel governance on environmental outcomes and capacity. There are simply too many variables to allow researchers to be able to isolate the effects of one variable from the other economic, political and institutional influences that shape policy. To be sure, at a theoretical level, one level of institutions may offer advantages over others. Local governance for

Table 2.1 An illustrative matching of environmental problems to governance levels

Governance Level	Chesapeake Bay	Urban Stormwater	Climate Change	Nanotech	Urban particulates	Sprawl	Chemical Products
Local	X	X	X	X	X	X	
Micro-regional	X	X	X		X	X	
State-provincial	X	X	X		X	X	X
Meso-regional	X		X		X		
National	X	X	X	X	X		X
Macro-regional			X		X		X
Global			X	X			X

Source: the author

Note: An 'X' indicates the author's assessment of governance levels that are substantially involved in responding to or managing a problem in the United States.

example, may usually be expected to be more responsive to community needs and preferences and provide more opportunities for citizen participation. At the same time, such discrete levels of governance cannot, on their own, respond to larger scale problems that exceed their authority and resources. At this point it becomes necessary to engage higher governance levels or to create problem or issue specific regional governance capacity to manage problems.

In the US, national environmental governance was strengthened in the 1970s for many reasons. One was the fear that states would compete to attract economic development with the promise of lax pollution controls, in the proverbial 'race to the bottom.' Another reason was that national institutions were seen as being more effective in standing up to powerful economic interests. Other reasons for creating national governance capacity through such laws as the Clean Air Act and Clean Water Act were to provide citizens with a basic level of environmental and health protection (such as through the National Ambient Air Quality Standards or Safe Drinking Water Act Primary Standards); to enjoy the economies of scale from national level standard setting, research and technology development; and to draw upon the superior fiscal resources of the national government for such purposes as building critical environmental infrastructure (such as sewage treatment plants) and cleaning up abandoned waste sites. Since then the lines between national and state authority in the US have been a recurring topic of debate.

Indeed, it is reasonable to argue that many states now are more innovative and serious about environmental policy and performance than is the federal government (Rabe 2010). Climate change is the leading example. Regional carbon dioxide trading programs pulled together by many state governments have been formed in the US, most notably the Regional Greenhouse Gas Initiative in the eastern part of the country (see Rabe 2004; Betsill and Rabe 2009). More than half of the states have adopted Renewable Portfolio Standards that require utilities to generate a specified percentage of their power from renewable sources. State and local sustainability and climate change planning greatly exceeds what is occurring at the national level (Portney 2003; Slavin 2011). These indicators of state and local leadership suggest that the learning and diffusion effects that have been observed in the EU could play a prominent role in the evolution of US environmental policy in coming decades, as the states compensate for policy gridlock on sustainability and climate issues at the national level.

Several concluding points may serve to pull this discussion together and set the stage for the rest of the volume. First, although it is not feasible to draw general conclusions about the effects of patterns of multilevel governance, it may be worth thinking about governance levels as part of a larger institutional configuration that can influence policy capacity. This is especially important as governance at all levels undergoes a transition from environmental protection as the dominant policy model and discourse to an era of environmental sustainability (Mazmanian and Kraft 2009; Fiorino 2001; Fiorino 2010). Some of the institutional research reviewed above suggests that systems with a greater capacity to integrate across policy sectors and interests are more successful in achieving high levels of environmental performance. Much of the qualitative research reaches similar conclusions. Do the patterns in multilevel governance affect this integrative capacity? Are there institutional characteristics at different levels of governance that will be more important in an era of environmental sustainability than in one of environmental protection? What are the roles of governance levels in a sustainability era?

A second point to consider from this discussion is the relationships among governance levels and the stages in the policy cycle. Organizing research more explicitly on the basis of the stage of policy making may open a window to achieving a greater understanding of the effects of multilevel governance. The stages of agenda setting, formulation, legitimation, implementation and evaluation/change vary in terms of the institutions that are involved, the role of analysis and information, the dynamics of the interactions among key actors, and even in the relationships among governance levels themselves. A promising area for future research would be to develop an understanding of the effects of multilevel governance at different stages of the policy cycle.

A third conclusion suggested from this discussion is that the direction of overall effects is likely to be neutral. That is, having strong multilevel institutions is no more likely to promote environmental protection and sustainability than it is to impede it. The studies cited in this paper do not attribute a strong independent effect to either strong or weak multilevel institutions. Committed national leadership in an institutional setting that favours collective action in support of environmental values is more likely to result in greener policies in a more unitary system. The opposite could be true as well. Had the US had a strong unitary system from 2001–9, there would have been even less progress towards climate action; if it had a strong unitary system now, there could well be more progress. Different vertical and horizontal configurations of power tend to favour certain interests in some situations but not others. It is difficult to draw general conclusions based on the evidence available now. It may be possible, however, to generalize about the effects of institutional characteristics on the ability to implement and sustain policy change. The stages of agenda setting, policy formulation, implementation and evaluation/revision are a place to start in such an institutional analysis.

A fourth point is the importance of learning more about the best policy fit between problems and governance levels. Are some governance levels more suited to resolving some kinds of problem than others? The existing assignment of authority and roles to different levels reflects a combination of political, economic and other influences over time. For some issues the basis for allocating governance authority is clear. Problems requiring regulation of products, for example, tend to be resolved at higher levels. The pattern is to push such issues as chemicals or vehicle regulation to higher levels, because products are sold in national and global markets. By contrast, for problems linked strongly to a geographic area (such as urban sprawl), authority lies largely with local governments, which manage land use and zoning, transportation and other such development issues. However, even for these, higher levels of governance affect the results. National and state policies affect various aspects of energy use and costs, transportation funding, building standards, and brownfields remediation, all of which influence local development.

Related to this issue of the best fit between problems and governance levels is the relationship with different categories of policy tools. For example, it is generally argued that air emission trading programs are more likely to be cost effective if they cover a larger area and a greater number of sources. So it makes sense to apply them at higher levels of governance. Public-private partnerships on the other hand, may be more manageable and effective locally, where there may be greater social capital and a basis for building better collaborative relationships. Are some policy tools and strategies more effective at different governance levels? Is it possible to offer general propositions about the three-way fit among governance levels, policy tools and the characteristics of environmental problems? These are promising areas for future research.

In summary, this chapter recommends four general questions for research and policy on multilevel governance, examining:

1. How different patterns of multilevel governance interact with other explanations for national and subnational environmental performance.
2. How the effects of multilevel governance patterns vary at stages of the policy cycle.
3. Whether some governance patterns are more effective at dealing with particular problems.
4. Understanding the implications of a transition from an era of environmental protection to one of environmental sustainability for the effective engagement of governance levels.

REFERENCES

Barrett, S. and K. Graddy (2000), 'Freedom, growth, and the environment,' *Environment and Development Economics* **5**, pp. 433–56.

Battig, M.B. and T. Bernauer (2009), 'National institutions and global public goods: Are democracies more cooperative in climate change policy?' *International Organization* **63**, 281–308.

Bernauer, T. and V. Koubi (2009), 'Effects of political institutions on air quality,' *Ecological Economics* **68**, 1355–65.

Betsill, M. and B.G. Rabe (2009), 'Climate change and multilevel governance: The evolving state and local roles,' in D.A. Mazmanian and M.E. Kraft, *Toward Sustainable Communities: Transition and Transformations in Environmental Policy*, Cambridge, MA: MIT Press, pp. 201–25.

Börzel, T.A. (2000), 'Why there is no "southern problem": On environmental leaders and laggards in the European Union,' *Journal of European Public Policy* **7**, 141–62.

Cole, M.A., A.J. Rayner, and J.M. Bates (1997), 'The environmental Kuznets curve: An empirical analysis,' *Environment and Development Economics* **2**, 401–16.

Crepaz, M.L. (1995), 'Exploring national variations of air pollution levels: Political institutions and their impact on environmental policy-making,' *Environmental Politics* **4**, 391–414.
Dasgupta, S., A. Mody, S. Roy, and D. Wheeler (2001), 'Environmental regulation and development: A cross-country empirical analysis,' *Oxford Development Studies* **29**, 173–87.
Dasgupta, S., B. Laplante, H. Wang, and D. Wheeler (2005), 'Confronting the environmental Kuznets curve,' in R. Stavins (ed.), *Economics of the Environment: Selected Readings*, 5th ed. New York: W.W. Norton, pp. 399–422.
Dasgupta, S., K. Hamilton, K.D. Pandy, and D. Wheeler (2006), 'Environment during growth: Accounting for governance and vulnerability,' *World Development* **34**, 1597–1611.
Dinda, S. (2004), 'Environmental Kuznets curve hypothesis: A survey,' *Ecological Economics* **49**, 431–55.
Dryzek, J.S. (1997), *The Politics of the Earth: Environmental Discourses,* Oxford: Oxford University Press.
Duwel, A. (2010), 'Democracy and the environment: The visibility factor.' Available as a working paper at SSRN: http://ssrn.com/abstract=1582299.
Emerson, J., D.C. Esty, M.A. Levy, C.H. Kim, V. Mara, A. de Sherbinin, and T. Srebotnjak (2010), *2010 Environmental Performance Index,* New Haven, CT: Center for Environmental Policy.
Esty, D.C. and M.E. Porter (2005), 'National environmental performance: An empirical analysis of policy results and determinants,' *Environment and Development Economics* **10**, 391–434.
Fiorino, D.J. (2001), 'Environmental policy as learning: A new look at and old landscape,' *Public Administration Review* **61**, 322–34.
Fiorino, D.J. (2010), 'Sustainability as a conceptual focus for public administration,' *Public Administration Review* **70S**, S78–88.
Fiorino, D.J. (2011), 'Explaining national environmental performance: Approaches, evidence, and implications,' *Policy Sciences* **44**, 367–89.
Frederiksson, P.G. and D.L. Millimet (2004), 'Electoral rules and environmental policy,' *Economics Letters* **84**, 237–44.
Frederiksson, P.G. and J.R. Wollscheid (2007), 'Democratic institutions versus autocratic regimes: The case of environmental policy,' *Public Choice* **130**, 381–93.
Gallagher, K.P. and S.C. Thacker (2008), 'Democracy, income, and environmental quality,' *Working Paper* 164, Political Economy Research Institute, University of Massachusetts.
Grossman, G. and A.B. Krueger (1995), 'Economic growth and the environment,' *The Quarterly Journal of Economics* **110**, 353–77.
Harrison, K. and L.M. Sundstrom, (eds) (2010), *Global Commons, Domestic Decisions: The Comparative Politics of Climate Change,* Cambridge, MA: MIT Press.
Holzinger, K., C. Knill, and T. Sommers (2011), 'Is there convergence of national environmental policies? An analysis of policy outputs in 24 OECD countries,' *Environmental Politics* **20**, 20–41.
Jahn, D. (1998), 'Environmental performance and policy regimes: Explaining variations in 18 OECD countries,' *Policy Sciences* **31**, 107–31.

Jänicke, M. (1997), 'The political system's capacity for environmental policy,' in M. Jänicke and H. Weidner (eds), *National Environmental Policies,* New York: Springer, pp. 1–24.

Jänicke, M. (1992), 'Conditions for environmental policy success: An international comparison,' *The Environmentalist* **12**, 47–58.

Jänicke, M. and K. Jacob (2004), 'Lead markets for environmental innovations: A new role for the nation state,' *Global Environmental Politics* **4**, 29–46.

Jänicke, M. and H. Weidner (eds) (1997), *National Environmental Policies: A Comparative Study of Capacity-Building,* New York: Springer.

John, D. (2004), 'Civic environmentalism,' in R.F. Durant, D.J. Fiorino, and R. O'Leary, (eds), *Environmental Governance Reconsidered: Challenges, Choices, and Opportunities*, Cambridge, MA: MIT Press, pp. 219–54.

Jörgens, H. (2004), 'Governance by diffusion: Implementing global norms through cross-national imitation and learning,' in W.M. Lafferty (ed.), *Governance for Sustainable Development: Adapting Form to Function,* Cheltenham, UK: Edward Elgar, pp. 246–83.

Knill, C., M. Debus, and S. Heichel (2010), 'Do parties matter in international policy arenas? The impact of political parties on environmental policy outputs in 18 OECD countries, 1970-2000,' *European Journal of Political Research* **49**, 301–36.

Kraft, M.E. and N.J. Vig (2010), 'Environmental policy over four decades: Achievements and new directions,' in Vig, N.J. and M.E. Kraft, *Environmental Policy: New Directions for the Twenty-First Century*, Washington, D.C.: CQ Press, pp. 1–26.

Lafferty, W.M. and J. Meadowcroft (eds) (2000), *Implementing Sustainable Development: Strategies and Initiatives in High-Consumption Societies*, Oxford: Oxford University Press.

Levinson, A. (2002), 'The ups and downs of the environmental Kuznets curve,' in J.A. List and A. de Zeeuw (eds), *Recent Advances in Environmental Economics*, Cheltenham, UK: Edward Elgar, pp. 119–41.

Lenschow, A., Liefferink, D., and S. Veenman (2005), 'When the birds sing: A framework for analyzing domestic factors behind policy convergence,' *Journal of European Public Policy* **12**, 797–816.

Li, Q. and R. Reuveny (2006), 'Democracy and environmental degradation,' *International Studies Quarterly* **50**, 935–56.

Liefferink, D., B. Arts, J. Kamstra, and J. Ooijevaar (2009), 'Leaders and laggards in environmental policy: A quantitative analysis of domestic policy outputs,' *Journal of European Public Policy* **16**, 677–700.

Mazmanian, D.A. and M.E. Kraft (2009), 'The three epochs of the environmental movement,' in D.A. Mazmanian and M.E. Kraft, *Toward Sustainable Communities: Transition and Transformations in Environmental Policy*, Cambridge, MA: MIT Press, pp. 3–32.

Miah, M.D., M.F.H. Masum, M. Koike, S. Akther, and N. Muhammed (2011), 'Environmental Kuznets curve: The case of Bangladesh for waste emission and suspended particulate matter,' *The Environmentalist* **31**, 59–66.

Munasinghe, M. (1999), 'Is environmental degradation an inevitable consequence of economic growth: Tunneling through the environmental Kuznets curve,' *Ecological Economics* **29**, 89–109.

National Academy of Public Administration (2007), *Taking Environmental Protection to the Next Level: An Assessment of the U.S. Environmental Services Delivery System*, Washington, DC.
Neumayer, E. (2002), 'Do democracies exhibit stronger environmental commitment? A cross-country analysis,' *Journal of Peace Research* **39**, 139–64.
Payne, R. (1995), 'Freedom and the environment,' *Journal of Democracy* **6**, 41–55.
Perrings, C. and A. Ansuategi (2000), 'Sustainability, growth, and development,' *Journal of Economic Studies* **27**, 19–54.
Poloni-Staudinger, L.M. (2008), 'Are consensus democracies more environmentally effective?' *Environmental Politics* **17**, 410–30.
Portney, K.E. (2003), *Taking Sustainable Cities Seriously: Economic Development, the Environment, and Quality of Life in American Cities*, Cambridge, MA: MIT Press.
Rabe, B.G. (1999), 'Federalism and Entrepreneurship: Explaining American and Canadian Innovation in Pollution Prevention and Regulatory Innovation,' *Policy Studies Journal* **27**, 209–20.
Rabe, B.G. (2004), *Statehouse and Greenhouse: The Emerging Politics of American Climate Change Policy*, Washington, DC: Brookings Institution.
Rabe, B.G. (2010), 'Racing to the top, the bottom, or the middle of the pack? The evolving state government role in environmental protection,' in N.J. Vig and M.E. Kraft, (eds), *Environmental Policy: New Directions for the Twenty-First Century,* Washington, DC: CQ Press, pp. 27–50.
Rabe, B.G., and M. Gaden (2009), 'Sustainability in a regional context: The case of the Great Lakes Basin,' in D.A. Mazmanian and M.E. Kraft, *Toward Sustainable Communities: Transition and Transformations in Environmental Policy*, Cambridge, MA: MIT Press, pp. 289–314.
Recchia, S.P. (2002), 'International environmental treaty engagement in 19 democracies,' *Policy Studies Journal* **30**, 470–94.
Scruggs, L. (2003), *Sustaining Abundance: Environmental Performance in Industrial Democracies*, Cambridge, UK: Cambridge University Press.
Slavin, M.I., (ed.) (2011), *Sustainability in America's Cities: Creating the Green Metropolis,* Washington, DC: Island Press.
Stern, D.I. (2004), 'The rise and fall of the environmental Kuznets curve,' *World Development* **32**, 1419–39.
Vogel, D. (1993), 'Representing diffuse interests in environmental policymaking,' in R.K. Weaver and B.A. Rockman (eds), *Do Institutions Matter? Government Capacities in the United States and Abroad*, Washington, DC: Brookings Institution, pp. 237–71.
Vogel, D. (1997), 'Trading up and governing across: Transnational governance and environmental protection,' *Journal of European Public Policy* **4**, 556–71.
Wälti, S. (2004), 'How multilevel structures affect environmental policy in industrial countries,' *European Journal of Political Research* **43**, 599–634.
Weidner, H. (2002), 'Capacity-building for ecological modernization: Lessons from cross-national research,' *American Behavioral Scientist* **45**, 1340–68.

Weidner, H. and M. Jänicke, (eds) (2002), *Capacity Building in National Environmental Policy: A Comparative Study of 17 Countries*, New York: Springer.

PART II

Multilevel governance of water resources

3. Subsidiarity as a 'scaling device' in environmental governance: the case of the European Union

David Benson and Andrew Jordan

Identifying the appropriate balance between the collective need for higher-level intervention and the maintenance of lower level autonomy in policy making, is an inherent problem within all multilevelled systems of governance.[1] It is a particularly important issue in relation to European Union (EU) environmental policy making, where determining the most appropriate level of governance to address policy problems raises many thorny questions relating *inter alia* to sovereignty, democratic legitimacy and economic costs. The EU's experience demonstrates that 'scale matters' (Young 2002, 27). Indeed it is an issue that EU policy makers have been grappling with for nearly four decades. 'Scale' primarily concerns 'the levels at which phenomena occur in the dimensions of time and space' (Young 2002, 26). Therefore, defining the scale at which particular environmental problems occur is an important factor in deciding how they will be addressed by policy systems. When governance spans, as it does in the EU, many levels, policy making 'tasks' will also need to be spread across a political space encompassing local, regional, national and supranational institutional actors (Weale 2002, 348). Logically it may be more effective and efficient to address 'local' issues at local levels, and Europe-wide issues supranationally at a European level (Weale, 2002). But, in practice, environmental problems invariably have complex multiscale characteristics that render such generalizations problematic and hence are productive of political controversy. In this chapter we argue that the EU offers a very useful context in which to examine how scale issues are perceived and addressed by policy makers.

Throughout much of its recent history, the EU has sought to resolve these dilemmas by employing the federal theoretical principle of subsidiarity. This essentially suggests that problems should be addressed at the lowest effective level commensurate with their scale 'unless allocating them to a higher-level central unit would ensure higher comparative efficiency or effectiveness in achieving them' (Follesdal 1998, 190). Since its inclusion in the 1986 Single European Act (SEA), subsidiarity has provided a principle for the allocation of policy making tasks between the EU, Member States and lower levels. Crucially, in the EU, environmental policy is neither the sole responsibility of the EU institutions nor the 28 Member States; rather, it is a *de jure* area of shared powers. Essentially, the subsidiarity principle seeks to allocate tasks among these levels according to the scale characteristics of particular problems, i.e. it is the EU's preferred 'scaling device.' Studying the EU's application of subsidiarity could therefore reveal much about how far tasks could be 'scaled up' or 'scaled down' between levels. In this respect, the experiences of the EU (with some caveats – see below) could also provide valuable lessons for designing (multilevel) governance systems more generally.

Before proceeding we would like to sound one important note of caution. Academics have consistently pointed to the unique, or *sui generis*, nature of European integration. Within EU scholarship this has triggered a lively debate about whether the EU is in fact an 'n' of 1 (i.e. a single, unique case which is beyond compare) (Nyikos and Pollack 2003). While the EU is not a 'state' in the widely accepted sense, it possesses many more 'state-like' qualities than a standard trans-national governance regime (Jordan 2005). Indeed it remains a unique political hybrid: an economically based international organization, in which the 28 Member States have progressively shared certain policy making tasks with EU institutions. These differences notwithstanding, when viewed as a federal type system of multilevel governance, we believe that the EU can be fruitfully compared with other political systems such as the US, Australia and Canada (Burgess 2006; Hueglin and Fenna 2005; Menon and Schain 2006; Watts 2008; Weibust 2009; Benson and Jordan 2010a), and consequently presents a fascinating and potentially rewarding case in which to study scale issues.

The chapter is structured around several interrelated questions regarding the EU's deployment of the subsidiarity principle. First, what is meant by the principle of subsidiarity? Here we define subsidiarity and show how competing theoretical interpretations have characterized its development and implementation. Second, why was there a need for scaling in EU governance? We examine the historical context to the introduction of subsidiarity and its ongoing application. Third, how has subsidiarity been used as a scaling device within environmental policy and what have been its actual effects? Here the practical application of subsidiarity is examined with a focus on one policy area in which tasks have been continually contested, namely the water quality sector. Fourth, how successful has subsidiarity been as a scaling device in terms of environmental governance? We then analyse how scaling has influenced environmental outcomes on the ground. Finally, what can be learnt from the EU's experience with subsidiarity for other systems of multilevel governance? Some pointers are provided towards understanding the circumstances under which scaling might (not) be successful in addressing environmental policy problems. We also offer some conclusions in relation to the development of a more general 'theory of scale' for multilevel governance.

THE PRINCIPLE OF SUBSIDIARITY

In the EU, the principle of subsidiarity has emerged as the primary *de jure* mechanism for allocating tasks in areas (like environmental protection) of shared powers. It arguably provides a strong presumption in favour of lower level autonomy (Jordan and Jeppesen 2000, 66). But while its general meaning is well understood, its practical application is open to interpretation and thus political contestation.

Part of this situation arises because its philosophical roots are deep and very complex. Consequently, multiple competing conceptions of subsidiarity have emerged within political theory to forward arguments on the effectiveness, efficiency and legitimacy of scaling (for an overview, see Follesdal 1998). While the problem of how to allocate tasks can be found in the work of ancient scholars like Aristotle, issues of scaling were first addressed in depth by Johannes Althusius in the seventeenth century (Follesdal 1998). Based on Calvinist doctrines, his political theory of federalism argued for the location of sovereignty at the lowest possible political level, i.e. with territorial 'sub-units' incorporating families and communities, to stop authorities from intruding too much into personal beliefs and activities (Follesdal 1998, 200–203). Since Althusius penned his arguments, theories have been developed to encompass many differing normative interpretations of subsidiarity.

These normative precepts can be characterized in several ways but broadly they provide arguments in favour of the centralization and/or decentralization of tasks. From an Althusian perspective, tasks should normatively be scaled to the lowest level possible since this is perceived as being more effective in solving problems in the most efficient way while best serving the interests of citizens. These types of ideas have also come to underpin liberal perceptions of subsidiarity, whereby the principle serves to prevent unnecessary state intrusion into individual freedoms, i.e. a 'single-edged sword' (Jordan and Jeppesen 2000, 66). For example, liberal contractualist theory argues that subsidiarity should be based on limited higher-level intervention in lower-level decision making 'allowing local accommodation both to circumstances and to expectations' (Follesdal 1998, 211). By contrast Catholic social theory normatively views the concept more as a 'double-edged' sword (Jordan and Jeppesen 2000, 66), According to Burgess (2000, 225) it emerged from a 'peculiarly Catholic view of organic society' as articulated in several papal encyclicals. In a reaction to political liberalism and its promotion of self-interest, Catholic social thought sought to promote the state as the protector of the common good within capitalist societies. Subsidiarity in this sense seeks to protect individual liberties and lower level tasks but also allows state intervention where more effective or efficient in achieving the common good. This 'Janus-faced' feature of subsidiarity was to prove highly attractive to EU policy makers as they struggled with defining and resolving issues of scale, something that we explore in the next section (Golub 1996, 686).

HISTORICAL ANTECEDENTS: SUBSIDIARITY AS A SCALING DEVICE

Disputes over the scaling of tasks in the EU are intrinsically linked to its origins and subsequent patterns of integration. In the period immediately after the formation of the EU's founding body, the European Coal and Steel Community (ECSC) in 1951, pioneering European elites unsuccessfully attempted to create two proto-federal organizations, namely the European Political Community and the European Defence Community. Their demise due to the overriding nationalist interests of states, led elites such as Jean Monnet to adopt a different approach to cementing European union (Duchêne 1994). His solution was not to attempt to create a potentially unstable federation but to cultivate limited integration between states in areas of mutual economic activity in order to create functional pressures for greater political cooperation: i.e. federal political integration would follow economic integration. Monnet's 'method' of functional scaling was formative in the subsequent creation and evolution of the European Economic Community (EEC) established by the Treaty of Rome in 1957. Here six European states pooled their sovereignty in certain limited policy areas, most notably agriculture and trade, as tasks were rescaled upwards to the European level.

However, the spectacular success of the Monnet Method in driving forward European integration led to rising tensions with Member States. In what scholars have described as 'integration by stealth' (Weale 2005; Majone 2009), the European Commission repeatedly sought to broaden the remit of the then EEC's tasks into cognate policy sectors as a multilevel system of governance gradually emerged, while eschewing any real debate over what Haas (1958) called 'task expansion.' This may have had the intended effect of binding Member States into a system of EU governance, but it also helped to generate a significant democratic and legitimacy deficit that haunts the EU to this day (Majone 2009). As the EEC incrementally expanded its tasks it eventually came into direct confrontation with governments anxious about the erosion of their national sovereignty. Matters came to a head in the mid-1960s when the French President de Gaulle temporarily withdrew his country from Community business in protest (Rosamond 2000). Settlement of the so called 'empty chair crisis' resulted in national governments securing more powers of veto in the Council of Ministers thereby ushering in greater intergovernmentalism (ibid). According to Hoffmann (1966), governments could therefore support scaling up in certain areas of low politics such as trade within the common market, to promote their national interests, but resisted incursions into their sovereign powers where they were threatened.

This feature was to an extent apparent as the EEC expanded its tasks into a new policy sector, namely the environment. When modern environmental policy making began to take off in the late 1960s, environmental problems were mostly addressed at either national or (more commonly) subnational levels (Jordan and Liefferink 2004, Chapter 3). Over time, the increasingly transboundary nature of many emerging environmental problems pushed national governments to adopt more international responses. However, disputes quickly arose between states, which resulted in demands for more coordinated responses at the European level. Problematically, the environment had not been explicitly mentioned in the list of EEC tasks or competences in the Treaty of Rome, even though some environmentally related legislation related to the functioning of the common market had been introduced (Jordan 2005). Heads of state, meeting at the Paris Council in 1972, then reacted to the growing clamour for action, and instructed the European Commission to prepare a policy for addressing environmental problems. It duly responded by publishing an Environmental Action Programme (EAP) in 1973 that laid out general principles and areas for policy development. Thereafter, the Commission embarked on the gradual creation of a *de facto* environmental policy, introducing measures for *inter alia* water quality, waste management, air pollution and wildlife protection. By the early 1980s a significant rescaling of tasks from the national to the European level was underway in the environmental policy sector, causing several high level disputes over the need for, and subsequent characteristics of, legislation. One notable example was the development of the 1988 Large Combustion Plant Directive (Weale et al. 2000). Acting in response to acidification of European forests and lakes, the Commission attempted to scale up tough standards for power station emissions based on domestic measures implemented by West Germany. These were actively resisted by laggard states such as the United Kingdom (UK) on the grounds of costs and sovereignty, as arguments over the legitimacy, effectiveness and efficiency of European level environmental tasks escalated. Environmental policy at this point became a touchstone for wider debates over the direction of integration, with states such as the UK favouring a deregulated, free trade based conception as opposed to others who envisaged deepening social and political union.

In order to assuage these fears, in the environmental policy sector at least, 'subsidiarity' was introduced as a key legal principle for task scaling as the Commission embarked on its plans for the single market. This objective entailed rescaling tasks in many areas, including the environment, in order to harmonize national standards. Under pressure from British government representatives, the Commission inserted a reference to the principle into the text of the 1986 Single European Act (Golub 1996, 689). Art. 130r(4) stated that:

> The [EU] shall take action relating to the environment to the extent to which the objectives … can be attained better at [EU] level than at the level of the individual member states (OJ L169, 29 June 1987)

Critically this article failed to slow down, let alone stop, the process of scaling up as the single market program was rolled out. It did not even stimulate a wide-ranging debate within the environmental sector on EU task allocation (Golub 1996). However, public opinion in Europe gradually started to turn against deeper integration (i.e. scaling up) more generally, meaning the issue did become more explicitly politicized (Jordan 2000, 1307).

Subsidiarity only came to wider prominence during the contentious negotiations over the Treaty on European Union in 1992. The draft of the Maastricht Treaty (as it is now commonly known) proposed scaling up some important policy areas for the first time. Pillar one of the draft treaty establishing the European Community included new areas such as consumer protection, while extending the EU's existing remit in others, for example, environment (Nugent 1994, 73). Under pillars two and three, the EU would assume additional powers for foreign policy, security policy, justice and home affairs. Member States' unease at the proposed scaling up led negotiators to move the subsidiarity principle from the environmental part of the Treaty to an introductory section, i.e. it would apply to all tasks.

But even at this point, differences of interpretation were painfully apparent. Jordan and Jeppesen (2000, 66) recount how the British government saw subsidiarity as a means of insulating national powers from further EU incursions, i.e. a 'single-edged sword'. Indeed Burgess (2000) notes how the market liberal Thatcher government perceived the principle as a means of mitigating the federalizing tendencies of the European Commission. By contrast Germany believed that it protected its pre-existing and constitutionally enshrined allocation of tasks to the regional level (or Länder). Hence, in theory, subsidiarity placed an overt emphasis on greater lower level policy making. Art. A of the Treaty thus referred to 'creating an ever closer union among the peoples of Europe, in which decisions are taken as closely as possible to the citizen' (OJ C191, 29 July 1992). Within the Commission headed by Jacques Delors, subsidiarity was viewed in very different terms. Due to his philosophical roots in Catholic personalism (van Kersbergen and Verbeek 1994; Burgess 2000), Delors understood subsidiarity as a means of justifying intervention for the common good, i.e. a 'double-edged sword'. Additionally, and perhaps more significantly, it provided him with a convenient political fix – helping the EU square the circle of promoting deeper integration alongside meeting demands for national autonomy. Subsidiarity therefore emerged as something of a fig leaf to cover critical disputes that were confronting the EU; its 'all things to everyone' conceptual malleability allowing broad agreement to be reached between divergent actors on the direction of the integration project.

More explicit guidance on applying subsidiarity was provided by art. 3b(2) Title 1 of the Treaty, which declared that:

> In areas which do not fall within its exclusive competence, the [EU] shall take action, in accordance with the principle of subsidiarity, only if and so far as the objective cannot be sufficiently achieved by the Member States and can therefore, by reason of the scale or effects of the proposed action, be better achieved by the [EU] (OJ C191, 29 July 1992)

With regards to scaling up, this article identified the need for its 'necessity' (Jordan and Jeppesen 2000, 67). Additionally art. 3b(3) argued the means employed should not go beyond what is necessary to achieve the objective – the so-called proportionality principle. Since directives and regulations have 'direct effect'[2] in national laws, but recommendations are non-binding, instrument choice is relevant to the application of subsidiarity and thus the resolution of the vexed question of scale. The conception of subsidiarity provided by art. 3b, it is argued, was primarily 'negative' since it seemingly 'placed the burden of proof' for intervention on the EU, rather than forcing Member States to provide reasons for retaining tasks (Jordan and Jeppesen 2000, 67). As art. 3b implicitly endorsed Member State tasks in areas of shared powers, its official purpose was arguably to restrict further scaling up.

Beyond Maastricht, various steps were taken to refine the meaning more fully and steer the application of the subsidiarity principle. Guidelines were established at the 1992 European Council meeting in Edinburgh. These stated that the EU should only scale up under specific conditions, that is: where issues are transnational; where their effects would interfere with Treaty objectives; and where it would show 'clear benefits' (European Council 1992, 15). In balancing these powers, EU measures 'should leave as much scope for national decisions as possible' (ibid). The Commission also committed itself to justifying new proposals on the basis of subsidiarity and publishing annual reports on its application. The first of these was published in 1993, as part of a process that was eventually named 'Better Law-making'.

Even more precise criteria to guide task allocation were provided in a protocol attached to the 1997 Treaty of Amsterdam. Art. 5 stated that:

> For [EU] action to be justified, both aspects of the subsidiarity principle shall be met: the objectives of the proposed action cannot be sufficiently achieved by Member States … and can therefore be better achieved by action on the part of the [EU] (OJ C 340, 10 November 1997)

Other provisions included detailed descriptions of the requirements for EU action and the need for greater Commission consultation prior to it proposing legislation (Articles 5–12).

Further changes to the application of subsidiarity have occurred since 2000. Increasing concerns over the EU's perceived 'democratic deficit' provided a stimulus for changes (Follesdal and Hix 2006). In its 2001 White Paper on European Governance, the Commission suggested amending the EU Treaties to define clearer principles for task allocation (CEC 2001). These issues were subsequently raised in the 2001 Laeken Council Declaration, which initiated a process aimed at agreeing a European constitution. In the Constitutional Treaty subsequently developed, subsidiarity was again endorsed as *the* primary principle governing task allocation. While the Constitutional Treaty was dealt a fatal blow in 2005 by its rejection in Dutch and French national referenda, subsidiarity was still retained as the EU's primary legal scaling device in the following Lisbon Treaty, finally ratified in 2009. National governments were also given greater powers to veto legislation under a so-called 'yellow' and 'orange' card system, whereby they can act collectively to modify or prevent Commission proposals before they enter the policy process (Benson and Jordan 2010b). In parallel, the Commission renamed the 'Better Law-making' agenda 'Better Regulation' in the early 2000s (CEC 2005). As part of this change, major Commission proposals are required to include extended impact assessments, wider consultation and demonstrate compliance with the subsidiarity and proportionality principles. However, as we shall see in the next section, all these changes were not enough to prevent conflicts in the environmental policy sector and many tensions still exist over scaling.

SUBSIDIARITY IN PRACTICE: THE SCALING OF WATER TASKS

In the EU, the rationale for scaling up environmental tasks from one level (principally the national) to a higher (i.e. supranational) level stems from the innately multiscale characteristics of many problems. Given the high population densities and the relatively small size of many Member States, most environmental problems in the EU inevitably have cross-scale, transboundary features. Arguments about task scaling or 'what level should do what?' have therefore become inherent in EU environmental policy since its inception. Indeed the European Commission anticipated these conflicts in the First Environmental Action Programme in 1973, stating that:

> ... it is necessary to establish the level of action (local, regional and national, [EU], international) that befits the type of pollution and the geographical zone to be protected should be sought. Actions which are likely to be the most effective at [EU] level should be concentrated at that level; priorities should be determined with special care ... (OJ C112, 20 December 1973)

This test therefore provided a 'double-edged sword' conception that allowed for both EU and lower level action depending on a judgement of relative effectiveness. Despite this statement, the scaling up of tasks after 1973 involved limited explicit or consistent application of these criteria. Even the introduction of the subsidiarity principle in 1987 did little to solve conflicts over scaling environmental tasks, with many disputes erupting over this issue (Golub 1996; Jordan 2000; Weibust 2009). Nowhere are these tensions more apparent than in one critical sub-sector, namely water governance.

Water pollution was an issue identified as an area of priority action in the First Environmental Action Programme in 1973. By the early 1970s, Europe was enduring severe water pollution problems with rivers such as the Rhine acting as open sewers for industrial and human waste and many coastal waters contaminated by land-based sources. These water pollution problems were also often transboundary, providing the EU with both a rationale and justification for coordinating policy responses. In the period after the publication of the EAP, the Commission therefore gradually introduced proposals for harmonizing national legislation that led to directives on bathing water quality, dangerous substances in water, surface water for drinking, freshwater fisheries, shellfisheries, groundwater, drinking water and titanium dioxide pollution. These 'first generation' water measures were typically based on the Commission setting uniform environmental quality standards, which took little account of the assimilative capacity of the receiving environments or natural variability between states, nor did they harmonize product standards such as those for drinking water (Benson and Jordan 2008). Some governments, most notably the UK, voiced concerns over certain proposed measures such as the Bathing Water Directive (on the grounds of costs and effectiveness) and the Dangerous Substances in Water Directive (on the grounds of incompatibility with British management approaches), leading to the Commission changing the latter. Most governments, however, appeared supportive of these directives – at least until the implications of implementation became apparent.

Bitter battles then broke out between Member States and the Commission over the efficiency, effectiveness and legitimacy of water task scaling, most visibly with regards the Bathing Water Directive (Wurzel 2002; Benson and Jordan 2010a). As in several other items of water policy, its implementation was clouded by controversy. The UK government's initial response was to circumvent the requirements of the directive (which obliged Member States to identify coastal and inland sites used for swimming and monitor pollution levels so that they remained within certain limit values) by under-reporting bathing waters. While this action meant authorities avoided making investments in expensive water treatment plant to reduce pollution levels, it brought the UK into direct confrontation with a Commission determined to enforce correct implementation. Finally, under threat of Commission legal action, the UK government announced a major investment program in sewage treatment infrastructure. Buoyed by this success, the Commission pursued enforcement proceedings against other recalcitrant Member States. Not surprisingly the directive was subsequently targeted for rescaling under the subsidiarity principle (Jordan 2000; Wurzel 2002; Benson and Jordan 2010a) – see below.

Such arguments over scaling did not subside after the introduction of subsidiarity in the SEA. The Commission published its Fourth Action Programme in 1987 in which it pledged to control the agricultural pollution of waters. Then, acting on the conclusions of a ministerial meeting on the future of EU water policy in Frankfurt in 1988, it agreed to introduce legislative proposals to reduce sewage pollution. These events led to the Commission producing two more water policy proposals: one for reducing nitrates pollution from agriculture and another requiring wastewater treatment for large urban areas. Measures contained in these 'second generation' policies threatened to be even more costly than the earlier water directives, further alienating more Euro-sceptic states such as the UK (Benson and Jordan 2008). Both directives were adopted after protracted negotiation in 1991 amid an atmosphere of rising political tension in the EU as governments bargained over the Maastricht Treaty. Inclusion of the subsidiarity principle in the agreement, as we saw above, did little to diffuse conflicts over scaling – a feature apparent in the subsequent development of water policy.

After the publication of the Edinburgh guidelines for implementing subsidiarity, national governments led by the UK and Germany took the opportunity to submit 'hitlists' of legislative measures for repatriation, i.e. downward rescaling (Jordan 2000). Ten environmental proposals were subsequently withdrawn under the subsidiarity principle, including those addressing zoo animal welfare and dumping of waste at sea, although some were practically dead anyway (Haigh 2004; Golub 1996; Benson and Jordan 2011). The Commission was also able to resist repatriation of existing water tasks targeted in the lists by citing their popularity with the public and promising to revise them (Jordan 1999). Ironically this process led to two of them, the Drinking Water Directive and Bathing Water Directive, being made more stringent. A proposal for revising the Bathing Water Directive was published in 1994, which, after several iterations, resulted in a new directive in 2006 (Wurzel 2002; Benson and Jordan 2010a). Rather than radically rescale tasks, the final revised measure arguably appears to impose even tougher implementation standards, scheduled for introduction in 2014 (Benson and Jordan 2010a). On reflection, one conclusion may be that subsidiarity was indeed employed more as a 'double-edged sword' since it ultimately failed to stem upward task scaling in water policy.

But subsidiarity did have lasting impacts in less obvious ways. Haigh (2004) notes that in the wake of the subsidiarity debate, where EU water legislation had come under sustained criticisms from governments for its fragmented nature and focus on harmonized environmental quality, there was a growing demand for a different approach. Attempting to deflect these criticisms, the Commission introduced another proposal for a directive on the ecological quality of surface waters but this was considered inadequate for addressing these concerns by Member States and the European Parliament. The Commission's solution was to propose radical change in the form of the Water Framework Directive (WFD). Initially proposed in 2007 and adopted by the EU in 2000, this 'third generation' water policy repeals or assimilates several outmoded 'first generation' directives and introduces an integrated ecosystems-based approach predicated on EU level objectives setting, regional river basin planning and national flexibility in implementation (Benson and Jordan 2008). Instead of uniform environmental quality standards, which proved hugely controversial among states, the Directive obliges them to achieve 'good ecological status' for surface waters by 2015, a measure which is set according to specific river basins. This approach also integrates with recent water policies, most notably the 2007 Floods Directive, the revised Bathing Water Directive and also the 2008 Marine Strategy Directive. Forming the environmental 'pillar' of an emerging EU maritime policy, the latter directive follows the approach adopted by the WFD by requiring the development of integrated regional marine plans and the attainment of 'good environmental status' for waters. A new more subtle conception of subsidiarity has therefore emerged in water policy that appears to mean shared standard setting and lower level implementation flexibility – a departure from the centralized standard setting that clouded the development of earlier water tasks. This conception appears to have been translated into other areas of environmental policy more generally, as the EU has increased its use of a variety of policy steering instruments including more 'framework' type directives and non-regulatory strategies.

EFFECTIVENESS IN PRACTICE

Although it has manifestly influenced debates over task scaling, what most concerns environmentalists is how the discussion or application of subsidiarity has materially affected environmental quality in Europe. From this perspective scaling up arguably has had both positive and negative effects.

In quantitative terms, scaling up certainly has been positive in terms of the volume of environmental legislation introduced. The EU environmental *acquis communautaire* now constitutes an impressive body of legal mechanisms for protecting the environment. By the late 2000s it numbered over 200 significant items, as well as several hundred supporting actions (see IEEP 2010). The influence of the *acquis* on Member State environmental protection strategies is profound. Approximately 80 per cent of national environmental regulation is often claimed to be inspired, or in some way influenced by, the EU (RCEP 1998, 7). Without scaling up, many Member States, (and particularly those in Southern and Eastern Europe) would probably not have achieved their current levels of environmental protection (Jordan and Liefferink 2004).

Within specific sectors there have also been measurable improvements in environmental quality across the EU. In water quality for example, indicators employed by the European Environment Agency show some signs of pollution reduction (EEA 2011). Figures show that overall levels of nutrients in European freshwaters have declined since 1992, primarily due to reductions in pollution from wastewater point sources, phosphates from detergents and agricultural waste (ibid). Average levels of nitrates for example, declined by 9 per cent between 1992 and 2008 (ibid). Similar improvements are visible in coastal bathing water quality where compliance levels with mandatory standards reached 95.6 per cent of registered sites in 2008 – up from 82.3 per cent in 1992 (ibid). Nutrient levels in marine waters are also declining, although figures vary between different ocean areas. One of the main drivers for increasing water quality has been more widespread wastewater treatment across the EU. Around 80 per cent of populations in Northern and Southern states are now serviced by some form of wastewater treatment, although this contrasts markedly with former communist bloc eastern states where the figure is much lower (ibid). Rates for tertiary treatment of wastewaters (the highest standard) are also increasing across Europe. It could therefore be inferred that subsidiarity has been successful as a scaling device in encouraging better environmental governance.

While some trends are positive, significant question marks remain over other dimensions of scaling up. First, while most Member States have accurately transposed EU environmental legislation into their national laws, the Commission has still had to initiate many infringement proceedings under Treaty Article 226. Second, practical implementation also remains extremely variable (Jordan 2002). The European Court of Justice has continually ruled against Member States for noncompliance, with water directives proving consistently problematic (for examples, see Kramer 2002). Third – partly because of transposition, implementation and enforcement weaknesses – specific environmental problems have proved persistently difficult to resolve. One example is nitrate pollution of groundwater. Despite measures such as the 1991 Nitrates Directive, nitrogen levels in groundwater remain stubbornly high (EEA 2010). Another is the Bathing Water Directive. Despite the improvements in coastal water quality, Member States had collectively still not reached the mandatory targets, set in the mid-1970s, when the directive was reviewed in 2009. The new and arguably tighter standards introduced under the amendments to the directive are unlikely to be achieved easily due to diffuse pollution from agriculture (Benson and Jordan 2010a). Finally, it is debatable whether subsidiarity has really resolved debates around the democratic legitimacy, effectiveness or economic efficiency of the EU's activities. In the 2000s, the debate (which first surfaced in the early 1990s – see above) over the social and economic impacts of scaling up has strengthened appreciably. This time it has arisen not under the banner of subsidiarity, but under that of economic competitiveness. In fact, it could be argued that the moves towards increasing lower level flexibility, discussed above, relate more to the demands of powerful economic interests backed by national governments. Initiatives such as the Better Regulation agenda, the increasing use of 'soft' rather than hard law and more open methods of coordination whereby Member States coordinate their activities bilaterally rather than via legal harmonization at EU level, all demonstrate that the EU is now rethinking its whole approach to governance (Jordan et al. 2003; Jordan and Schout 2006).

CONCLUSIONS: TOWARDS A THEORY OF SCALE?

The EU's use of subsidiarity as a scaling device sheds light on how a rapidly maturing system of multilevel environmental governance has addressed the vexed issue of scale. When faced with multi-faceted problems that span spatial scales, the EU started by scaling up tasks that were traditionally exercised at lower levels of governance. Increasing resistance from Member States and growing public concern then led to the introduction of subsidiarity as a scaling device. In theory, subsidiarity can be used to ensure that policies achieve a better correspondence with the scale at which problems manifest themselves. But in practice it has proved to be an essentially contested term which some political actors, not least the Commission, have employed to justify more scaling up. While this has arguably enhanced environmental protection, with some improvements apparent, it has not been uniformly successful: implementation remains problematic; pollution problems are proving stubbornly resistant; democratic legitimacy has suffered; and many powerful groups (not least industry) have often mobilized against it on the grounds of economic competitiveness.

What lessons can we draw from the EU's experience with subsidiarity for both comparative learning and a more general theory of scale? First, while subsidiarity is probably the most obvious scaling principle for multilevel systems, trying to look for and employ it in a rational-objective manner is likely to be deeply problematic, raising a whole host of irreducibly normative matters concerning the process, efficiency and effectiveness of policy interventions. The EU's experience of trying to resolve the vexed issue of scale is that subsidiarity, even after repeated attempts at definition, can be interpreted in a myriad different ways to suit the preferences of different actors.[3] This feature relates strongly to the normative theoretical underpinnings of subsidiarity. So while in theory subsidiarity was originally intended by some actors to be a 'single-edged sword' designed to protect the interests of lower levels, in practice it has served the interests of others as a 'double-edged sword' with a strong bias in favour of greater EU integration. As Follesdal (1998 191) observes, 'since it merely places the burden of proof upon those who would centralize, it can have a centripetal effect by providing warrant for such moves.'

Second, while subsidiarity offers a way to frame a debate about scale, in the *realpolitik* of everyday problem solving, it is not easily defined and thus does not generate simple answers. Debates about scale in the EU have not been 'value free' or apolitical. Very often, they have been founded on wholly different interests and perceptions of scale, for example Member States demands, global influences and external lobbying. Since 1990, the issue of scale has actually become *more,* not less, politicized in the EU. A clear lesson for those advocating neat theoretical solutions to scaling issues is therefore that they can easily create more conflict and mistrust, not less.

Third, objective interpretations of subsidiarity appear to assume that a fairly discrete zero sum allocation of tasks between levels is possible. In reality, policy problems rarely arrive in discrete bundles. Again, the history of bathing water policy since the early 1970s provides a case in point. Attempts to set bathing water standards at beaches quickly revealed that most sources of pollution were not properly regulated. Wastewater treatment standards (embedded in the 1991 Urban Wastewater Treatment Directive) were then painstakingly agreed. But it now seems as though these will be inadequate to help meet new bathing water standards, given the polluting effects of diffuse runoff from agricultural land. The point here is that some aspects of 'the' policy problem are international (the flow of tourists across borders looking for clean water to bathe in), some are national (wastewater treatment standards) and some are profoundly local (farming practices). Whether by accident or design, the EU has gradually been drawn in to addressing all of them. Policy making in a word, has evolved *incrementally*, without a comprehensive discussion (let alone resolution) of the scale issue.

Finally the EU's experience with subsidiarity does nonetheless suggest that scale conflicts could potentially be confronted, if not completely resolved, to provide institutional design solutions in multilevel environmental governance. In recent times, disputes over scale have been partially offset by encouraging greater lower level participation in EU level policy making (inclusivity) and enhancing local autonomy in the implementation of policy (flexibility). This feature is apparent in third generation measures such as the Water Framework Directive and the Marine Thematic Strategy. So while nowadays there is a tacit acceptance by most EU actors that the majority of environmental tasks are better addressed supranationally, attempts are being made to better involve lower levels of governance to improve policy ownership. In such a setting the application of subsidiarity could then involve less a divisive zero sum battle between institutional levels over scaling up or down, and more a constructive dialogue over the sharing of functional tasks. As policy makers struggle globally with questions of scale, the EU's experiences are worth exploring as they yield important lessons, but unfortunately no easy solutions.

NOTES

1. Governance means established patterns of rule without an overall ruler (Jordan and Schout 2006). Because the EU undertakes the sort of tasks that governments have normally addressed, but is not itself a government, it is often referred to as a system of multilevel governance (Bomberg and Stubb 2003, 9; Bache and Flinders 2004).
2. Under EU law, directives and regulations have 'direct effect' in Member State legislation, i.e. they must be implemented directly.
3. For an illuminating case study of how the 'scale' of something quite discrete like bathing water quality, can be framed in so many different ways, see Wurzel (2002, 179–254) or Benson and Jordan (2010a). Weibust (2009) also shows that an issue which appears highly localised, such as noise levels, can nonetheless be drawn into EU policy making.

ACKNOWLEDGEMENTS

The authors would like to thank the ESRC Rural Economy and Land Use (RELU) project (RES-229-25-00009-A) for supporting the writing of this chapter through its funding of our comparative research into water governance. Further support was received from the Leverhulme Trust (Major Research Fellowship: reference number F00204AR).

REFERENCES

Bache, I. and M. Flinders (eds) (2004), *Multi-Level Governance*, Oxford: Oxford University Press.

Benson, D. and A. Jordan (2008), 'Understanding task allocation in the European Union: Exploring the value of federal theory,' *Journal of European Public Policy* **15**(1), 1–20.

Benson, D. and A. Jordan (2010a), 'The scaling of water governance tasks: A comparative federal analysis of the European Union and Australia,' *Environmental Management* **46**(1), 1–16.

Benson, D. and A. Jordan (2010b), 'European Union environmental policy after the Lisbon Treaty: Plus ça change, plus c'est la même chose?' *Environmental Politics* **19**(3), 468–74.

Benson, D. and A. Jordan (2011), 'Exploring the "tool-kit" of EU integration theory: What role for cooperative federalism?' *Journal of European Integration* **33**(1), 1–17.

Bomberg, E. and A. Stubb (2003), *The European Union – How Does it Work?* Oxford: Oxford University Press.

Burgess, M. (2000), *Federalism and European Union: The Building of Europe, 1950-2000*, London: Routledge.

Burgess, M. (2006), *Comparative Federalism: Theory and Practice*, London: Routledge.

Commission of the European Communities (CEC) (2001), *European Governance: A White Paper COM (2001) 428 final*. Brussels: CEC.

Commission of the European Communities (CEC) (2005), *Better Regulation for Jobs and Growth in the European Union COM (2005) 97 final,* Brussels: CEC.

Duchêne, F. (1994), *Jean Monnet: The First Statesman of Interdependence,* New York: Norton.

European Environment Agency (EEA) (2011), *Indicators,* Copenhagen: EEA, accessed on 1 July 2012 at http://www.eea.europa.eu/themes/water/indicators.

European Council (1992), *The European Council, Edinburgh 11–12 December 1992,* Brussels: General Secretariat of the Council.

Follesdal, A. (1998), 'Survey Article: Subsidiarity,' *The Journal of Political Philosophy* **6**(2), 190–218.

Follesdal, A. and S. Hix (2006), 'Why there is a democratic deficit in the EU: A response to Majone and Moravcsik,' *Journal of Common Market Studies* **44**(3), 533–62.

Golub, J. (1996), 'Sovereignty and subsidiarity in EU environmental policy,' *Political Studies* XLIV, 686–703.

Haas, E. B. (1958), *The Uniting of Europe. Political, Social and Economic Forces: 1950-1957*, Stanford: Stanford University Press.

Haigh, N. (2004), *Manual of Environmental Policy*, London: Maney Publishing and the Institute for European Environmental Policy.

Hoffmann, S. (1966), 'Obstinate or obsolete? The fate of the nation state and the case of Western Europe,' *Daedalus* **95**(3), 862–915.

Hueglin, T. and A. Fenna (2005), *Comparative Federalism*, Peterborough ON: Broadview.

Institute of European Environmental Policy (IEEP) (2010), *Manual of Environmental Policy,* London: IEEP.

Jordan, A.J. (1999), 'European Community water standards: Locked in or watered down?' *Journal of Common Market Studies* **37**(1), 13–37.

Jordan, A.J. (2000), 'The politics of multilevel environmental governance,' *Environment and Planning A* **32**, 1307–24.

Jordan, A.J. (2002), 'The Implementation of EU Environmental Policy,' in A.J. Jordan (ed.), *Environmental Policy in the European Union*, London: Earthscan, pp. 301–28.

Jordan, A.J. (2005), 'Introduction: European Union environmental policy,' in A.J. Jordan (ed.), *Environmental Policy in the European Union* (2nd ed), London: Earthscan, pp. 1–15.

Jordan, A.J. and T. Jeppesen (2000), 'EU environmental policy: Adapting to the principle of subsidiarity?' *European Environment* **10**(2), 64–74.

Jordan, A.J., R. Wurzel and A. Zito (eds) (2003), *New Instruments of Environmental Governance*, London: Frank Cass.

Jordan, A.J. and D. Liefferink (eds) (2004), *Environmental Policy in Europe: The Europeanization of National Environmental Policy*, London: Routledge.

Jordan, A.J. and A. Schout (2006), *The Coordination of the European Union: Exploring the Capacities for Networked Governance*, Oxford: Oxford University Press.

Kramer, L. (2002), *Casebook on EU Environmental Law*, Oxford: Hart Publishing.

Majone, G. (2009), *Europe as the Would-be World Power: the EU at Fifty*, Cambridge: Cambridge University Press.

Menon, A. and M.A. Schain (2006), *Comparative Federalism,* Oxford: Oxford University Press.

Nugent, N. (1994), *The Government and Politics of the European Union* (3rd Edition), Basingstoke: Macmillan.

Nyikos, S.A. and M.A. Pollack (2003), *Researching the European Union: Qualitative and Quantitative Approaches,* paper prepared for the European Union Studies Association (EUSA) Biennial Conference 2003. Nashville, TN, 27-29 March 2003.

Official Journal of the European Union (OJ) (various issues). Luxembourg: Office for Official Publications of the European Union.

Royal Commission on Environmental Pollution (RCEP) (1998), *Twenty-first Report: Setting Environmental Standards, Cm 4053*, London: HMSO.

Rosamond, B. (2000), *Theories of European Integration*, Basingstoke: Palgrave.

van Kersbergen, K. and B. Verbeek (1994), 'The politics of subsidiarity in the European Union,' *Journal of Common Market Studies* **32**(2), 215–36.
Watts, R. (2008), *Comparing Federal Systems*, Kingston: Queens University.
Weale, A. (2002), 'European environmental policy by stealth,' in A.J. Jordan (ed.) *Environmental Policy in the European Union*, London: Earthscan, pp. 124–40.
Weale, A., G. Pridham, M. Cini, D. Konstadakopulos, M. Porter, and B. Flynn (2000), *Environmental Governance in Europe,* Oxford: Oxford University Press.
Weibust, I. (2009), *Green Leviathan,* Aldershot: Ashgate.
Wurzel, R.K.W. (2002), *Environmental Policy Making in Britain, Germany and the European Union*, Manchester: Manchester University Press.
Young, O.R. (2002), *The International Dimensions of Environmental Change: Fit, Interplay, and Scale,* Cambridge MA: MIT Press.

4. Multilevel governance and the politics of environmental water recoveries

B. Timothy Heinmiller

In the middle two quarters of the twentieth century, irrigated agriculture expanded rapidly in many parts of the world, including western Canada, southeastern Australia and the western United States (US). Much of this expansion was facilitated by government policies that directed public money towards large scale dam and irrigation project construction, as well as liberal water allocation policies that sought to get as much water as possible into productive agricultural use. By the early 1970s however, the political consensus that had supported the era of irrigation expansion was under significant challenge. The emerging environmental movement, in particular, successfully challenged the construction of many new damming projects and generally worked towards the protection and restoration of riverine environments in irrigation areas. One of their most important efforts in this regard has been their push for environmental water recoveries. Environmental water recoveries are government programs that take water originally allocated for irrigation and reallocate it for environmental protection purposes. While many environmentalists see these water recoveries as essential for creating a sustainable balance between nature and economic production, most irrigators regard them as a fundamental threat to their livelihoods and communities, and both sides have worked to advance their interests through the political process.

Due to the federal nature of Canada, Australia and the United States, irrigator-environmentalist conflicts over environmental water recoveries have played out in multilevel political institutions. In all three countries, the states/provinces have primary constitutional jurisdiction over the allocation of water within their borders, but the federal governments have become involved in water allocation issues to various extents. The country with the least federal involvement has been Canada where, for various constitutional and political reasons, water allocation policies have been developed almost exclusively by the provinces. In contrast, the federal governments of Australia and the United States have been extensively involved in water allocation issues and water allocation policies have been developed at both the state and federal levels.

The purpose of this chapter is to investigate the relevance and importance of this varying federal involvement in shaping the political contest between irrigators and environmentalists over environmental water recoveries. It does so through a comparison of Alberta, New South Wales and California, the largest irrigation economies in their respective countries. All of these jurisdictions have experienced protracted political struggles over environmental water recoveries, yet, as noted above, there has been substantial variance in the level of federal involvement in water allocation issues. This therefore, allows us to see the impact that varying federal involvement has had on the politics of environmental water recoveries.

As the reader will see, the evidence suggests that greater federal involvement has generally worked to the advantage of environmentalists and that the more federal governments have been involved in water allocation issues, the more success environmentalists have had in achieving environmental water recoveries. This conclusion is supportive of findings in other areas of American environmental policy and reinforces the importance of multilevel institutional design in shaping environmental policy outcomes.

IRRIGATION AND THE POLITICS OF WATER

In the Canadian Prairies, southeastern Australia, and the American West, irrigated agriculture is a central feature of rural economies and governments have played a major role in irrigation development. Government objectives varied in early irrigation development but typically related to domestic food security and rural economic development (Glenn 1999; Worster 1985; Lloyd 1988). Government support for irrigation was secured through closed policy networks comprised of farmers, rural legislators and water reclamation bureaucrats, described aptly as 'iron triangles' (Gottlieb 1988, 46-52). These networks dominated water policy in all three countries and the actors within them shared a common purpose in the expansion of irrigation, disagreeing on occasion about the distribution of state largesse. The result was the pursuit of policy objectives such as 'full development' and 'closer settlement,' which were used to justify massive allocations of public money from all levels of government to put large tracts of 'wasted' land and large volumes of 'unused' water into agricultural production.[1]

One of the key institutional factors underlying the success and perpetuation of the iron triangles, and the expansionary policies they produced, was the relative absence of jurisdictional constraints on government spending in all three federations. Expansionary policies focused mostly on the supply dimension of water management, undertaking the construction of dams, weirs, canals and other infrastructure that could make water available to farmers and thereby increase irrigation production. The construction of such projects required large government outlays, so the focus of intergovernmental relations during this period was primarily on the distribution of project costs. Importantly, the divisions of power in these federations placed few restrictions on federal government spending, allowing them to spend in areas well outside their legislative jurisdiction.[2] The states/provinces for the most part welcomed this funding and were complicit in its implementation, helping to perpetuate the iron triangles that dominated water policy.

However, starting in the 1960s, continuous irrigation expansion began to be viewed in a more critical light as environmental problems began to surface and expected economic returns did not materialize. During the expansionary era, governments in all three jurisdictions distributed water entitlements liberally, but eventually this led to overallocation problems. Typically, no more than 40–50 per cent of the annual median flow of any river should be allocated to consumptive uses such as irrigation, in order to maintain the security of water entitlements and the health of aquatic environments. Yet by the 1990s, allocations in all three jurisdictions substantially exceeded this: in southern Alberta, allocations ranged from 68 per cent of median annual flow on the Bow River to 118 per cent of median annual flow on the St Mary River. In New South Wales, allocation on the Murrumbidgee River, the largest river entirely within the state, was 166 per cent of annual median flow. And in California, the State Water Resources Control Board estimated that '… there are roughly three times as many paper claims on surface water as water available in the system' (Alberta Environment 2003, 12; Murray-Darling Basin Ministerial Council 1995, 8; Hanak 2003, 9).

As rivers became overexploited, salinity and pollution problems increased, fish stocks declined, and aquatic environments degraded. The large dams that had made this intense water development possible became a favourite target of the emerging environmentalist movement and disappointing economic returns from many irrigation projects led others to question the wisdom of further irrigation expansion. This was especially the case since more irrigation would mean more dam construction and most of the good dam sites in these jurisdictions had already been developed, making further dam construction very expensive with low marginal returns (Davidson 1969; Randall 1981; Reisner 1993).

As the benefits of irrigation were increasingly questioned, the iron triangles that had long dominated water policy in Canada, Australia and the United States began to disintegrate, replaced by more open policy networks with a more diverse membership. The most important new entrants in these policy networks were environmentalists and neoconservatives. Environmentalists opposed irrigation expansion because it would result in further alteration of natural watercourses, depletion of streamflows and pollution of water sources, putting already fragile wetlands, aquatic environments and wildlife at greater risk. Many neoconservatives, by contrast, objected to further expansion because it would be funded by deficit-laden governments who they believed should cut public spending not increase it, especially on large public projects offering dubious economic returns. As environmentalists and neoconservatives gained traction in these policy networks, the changes in water policy were substantial. Federal funding for water development projects in the US dried up during the Carter administration and declined considerably during the 1980s in Australia and the 1990s in Canada (Reisner 1993, Chapter 9; Mertha and Lowry 2006, 7–8; Glenn 1999, 22).

As support for further irrigation expansion declined and environmentalists increasingly defined the terms of political debate, the focus in water policy networks shifted to environmental concerns. For environmentalists, bringing large-scale dam construction to a halt was not enough to protect and restore riverine environments. Irrigation related water allocations continued to grow and riverine environments continued to decline. Consequently, environmentalists turned their attention to government water allocation policies and worked vigorously for water allocation reforms. The first step in this campaign was to work towards 'freezes' or 'caps' in irrigation water allocations, halting further increases in irrigation water use and reserving the remaining unallocated water for the environment. The second, and even more difficult step, was to work for environmental water recoveries, effectively taking water that had been committed to agricultural production and reallocating it to increase river flows and restore riverine habitats.

For irrigators, the environmentalists' targeting of their water allocations constituted a serious threat to their livelihoods and communities. Water that had been abundant and relatively costless during the expansionary era was now scarce and relatively costly (Randall 1981). Those with low priority entitlements or low value crops were particularly threatened in the new policy environment. Consequently irrigators have been very active in resisting water allocation reforms and have continued to be a formidable political force, using their long experience in dealing with government, their close connections with key government agencies and their geographical and electoral concentration to their advantage. Yet the irrigators have enjoyed nowhere near the political dominance they once had as the old iron triangles have disintegrated, public opinion has become more sensitive to environmental issues and urbanization and industrial growth have diminished the relative economic importance of agriculture.

WATER ALLOCATIONS AND FEDERALISM

Although irrigator-environmentalist conflicts have pervaded the politics of water allocation in all three jurisdictions, these conflicts have been played out in distinctive multilevel institutional settings.

In Canada, the regulation of water taking and water use is primarily centred at the provincial level, especially in Alberta, which jealously guards its jurisdiction over natural resources from federal intrusion. Section 109 of the Constitution Act 1867 grants the provinces proprietary rights over most 'lands, mines, minerals and royalties' within their borders, thereby granting them proprietary rights over water by virtue of the common law riparian rights[3] doctrine (Kennett 1991, 24). Initially Alberta and Saskatchewan were denied these rights when they became provinces in 1905, but this was rectified in the 1930 Natural Resources Agreements, which specifically mention water as a provincial jurisdiction (Percy 2005). For its part the federal government has jurisdiction over water-related fields in navigation and shipping (Constitution Act 1867 s. 91[10]), fisheries (s. 91[12]), and international relations (s. 132), giving it a significant role, but largely at the margins of provincial regulation (Kennett 1991). For instance, the federal government participates as an equal partner with the United States in the International Joint Commission (IJC), and has a longstanding agreement on the apportionment of the St Mary River, which flows into Alberta from Montana, but once the water crosses into Alberta its allocation and use is considered a provincial matter. Similarly, the federal government is a signatory to the intergovernmental Master Apportionment Agreement, dividing the Prairie rivers between the three Prairie provinces, but has virtually no authority in intraprovincial water allocation decisions (Figliuzzi 2002). Much of this is explained by the provinces' determination to keep Ottawa out of their jurisdiction over natural resources, a desire that is particularly strong in Alberta, which has a long history of conflict with the federal government over the control of other natural resources such as oil. In short, the federal government is present in Alberta water allocation issues, but largely in the background since the Alberta government closely guards its jurisdiction over intraprovincial water allocation issues.

In Australia, the division of powers in water allocation was quite similar to that in Canada until a series of reforms in the last two decades dramatically increased the role of the federal government. The division of powers entrenched in sections 51–52 of the Australian Constitution designated specific areas of jurisdiction to the Commonwealth (federal) government, leaving all nonspecified areas to the states. Since neither the ownership nor the regulation of public waters were included in sections 51–52, the regulation of water taking was left primarily to the states (Jaensch 1997, 48–9). Like its Canadian counterpart, the Commonwealth government was a participant in an important interstate water apportionment agreement, the River Murray Waters Agreement of 1914, but its role was confined to protecting and enhancing navigation, a water use that diminished in importance over time (Clark 1983). Starting in the early 1990s however, the Commonwealth became increasingly involved in water allocation issues as the states of the Murray-Darling Basin, including New South Wales, proved unable to cope with water scarcity problems on their own. Initially the Commonwealth participated as a relative equal with the states in a series of intergovernmental water initiatives, bringing its national authority and greater spending power to bear on water allocation issues. However in 2005, during a period of extreme drought, the Howard government passed the Water Act, giving it the authority – with state acquiescence – to take over much of the water allocation authority in the Murray-Darling Basin traditionally reserved for the states. Subsequent governments have remained committed to the objectives of the Water Act and the Commonwealth now plays a dramatically more important role in water allocation than it did in the past (Connell 2007; Quiggin 2010).

In the United States the institutions of water regulation have also undergone a significant transformation in the last four decades as the federal government has increasingly asserted its prerogative in the regulation of water takings. In the American West, the exploitation of water resources was often initiated in advance of any significant government presence, guided only by common law and judicially derived principles such as prior appropriation.[4] Eventually, the states started to catch up and began regulating water takings, without derogating these original water rights (Gillilan and Brown 1997 16–19). State jurisdiction in water regulation however, was based on little more than the states' assertion of authority in this area, and state water policy was subject to federal preemption under the US Constitution. This was confirmed in the seminal *Rio Grande Irrigation Company* case of 1899 which '... made clear that state-authorized water use must not interfere with federal rights to protect the flow of the stream and can be superseded by the exercise of federal powers over commerce and public land' (Getches 2001, 6). Nevertheless, since state legislation already occupied the field and preempting this legislation would have caused massive uncertainty and instability in western economies, the federal government used its power sparingly for many decades. Yet, it remained a potential venue for water policy reform, and the federal government has shown an increased willingness since the early 1970s to involve itself in intrastate water allocation issues (Getches 2001). This has created a multilevel institutional setting, in which both state and federal policy venues have been active, and have constituted viable options for irrigators and environmentalists in their water allocation struggles.

So, although irrigator-environmentalist conflicts have pervaded Alberta, New South Wales and California over the last four decades, the institutional terrain on which these political conflicts have been staged has varied substantially. In Alberta, water allocation has been almost exclusively a provincial issue with little federal involvement; in New South Wales, water allocation authority has been incrementally ceded from the state to the federal level, so that the Commonwealth now plays a dominant role in this area; and in California, both the state and federal levels have long been active in water allocation issues.

The evidence presented below suggests that these variations in institutional design have had a role in shaping water allocation policy outcomes. Generally, federal involvement in water allocation has worked to the advantage of environmentalists as they have enjoyed much more success in federal policy venues that in state/provincial ones. This finding echoes those of some previous studies of American environmental federalism such as Koontz's *Federalism in the Forest* (2002) and Layzer's *Natural Experiments* (2008). Both of these studies found that developmental interests tended to be advantaged in state level policy processes whereas environmental interests tended to be advantaged in federal level processes (Koontz 2002; Layzer 2008). All of these studies suggest that the design of multilevel institutions is very important because they establish institutional terrains that can advantage some sets of interests over others.

ENVIRONMENTAL WATER RECOVERIES IN ALBERTA

Because the Canadian government has very little involvement in intra-provincial water allocation issues, environmental water recoveries in southern Alberta have been pursued primarily at the provincial level, a policy venue that has been much more sympathetic to irrigators than environmentalists. Historically, electoral politics in Alberta have been characterized by long periods of single party rule and the current ruling party is the Progressive Conservatives, who have held power continuously since 1971. One important element of the Progressive Conservatives' electoral base has been the 'irrigation belt,' a band of rural seats in southern Alberta in which farming is the dominant economic activity. Since the mid-1970s, the irrigation belt has consistently voted Progressive Conservative and a series of Progressive Conservative governments has consistently looked out for irrigators' interests (Brownsey 2008). A concerted environmentalist movement emerged in the Alberta water policy community in the mid-1980s, but the environmentalists have rarely been able to match the irrigators' influence in the Alberta government (Glenn 1999). The environmentalists do not form the same kind of geographically concentrated and electorally influential constituency that the irrigators do, nor are they part of the Progressive Conservative Party's electoral base. Confined to the provincial level and marginalized in the perennial governing party, environmentalists have been forced to pursue environmental water recoveries on the irrigators' home turf and as a consequence have enjoyed relatively little success thus far.

The rise of the environmentalist movement as a political force in Alberta water politics can be traced to the controversy surrounding the construction of the Oldman Dam. In 1984, the Alberta government decided to proceed with a dam on the Oldman River, a project that had been contemplated since the early 1970s and was designed to support irrigation expansion in the area. The decision sparked unprecedented resistance from local First Nations populations and served as the catalyst to unite disparate dam opponents into a concerted environmentalist movement (Glenn 1999). Although the environmentalists failed in their bid to stop the Oldman Dam, they gained recognition as legitimate stakeholders in the Alberta water policy community and succeeded in prodding the Alberta government to review its existing water allocation legislation, the 1930 Water Resources Act. Alberta Environment, the provincial environmental department, held public and stakeholder consultations on reforming the province's water rules between 1990 and

1996, the process eventually culminating in the passage of the new Alberta Water Act in 1999 (Bankes and Kwasniak 2005, 5).

Despite the new legislation, the irrigators and their water entitlements remained strongly protected. During the development of the Water Act, environmentalists pushed for legislative provisions to make the protection of riverine environments a priority over agricultural water uses. In overallocated river basins this implied that some water would be recovered from irrigators, probably from low priority irrigators with relatively recent entitlements and reallocated for purposes of environmental protection.[5] The irrigators understandably objected to any legislative provisions that threatened the security of their existing water entitlements. When provisions prioritizing environmental protection were included in the first draft of the new Water Act, introduced in the legislature in February 1996, agricultural interests responded by rallying the support of rural legislators, the Minister of Agriculture and senior Alberta Agriculture officials. Eventually they succeeded in having the Water Act changed to guarantee the sanctity of existing water licences, regardless of the environmental objectives established in other parts of the legislation (Smith 1996, 14–15; Percy 1996–7, 229). This was a major legislative victory for irrigators because it effectively preempted any significant attempt to take water from irrigators and reallocate it to environmental protection.

In the end the Water Act became a piece of legislation that implicitly recognizes a need for environmental water recoveries, but provides little means for achieving them. The Act calls for the creation of basin water management plans for each of southern Alberta's four sub-basins and as part of this planning process, the establishment of water conservation objectives for each sub-basin. In the three sub-basins that were already overallocated, the purpose of these provisions was to halt and reverse the environmental decline associated with overallocation, implying the necessity of environmental water recoveries. However the only part of the Water Act providing for any environmental water recoveries is the conservation holdback mechanism, and only in modest amounts. The Water Act also provided for the introduction of water licence trading in southern Alberta and the conservation holdback mechanism allows Alberta Environment to withhold up to 10 per cent of the volume of any water traded for the purposes of protecting the aquatic environment or meeting a water conservation objective (Percy 1996–7, 239). So when existing water licences are traded, a small portion of the licensed water, which was probably used for irrigation, can be recovered and reallocated to the environment.

Although clearly a means of recovering environmental water, the amount of water recovered through conservation holdbacks has been insubstantial. As of May 2010, only 28 conservation holdbacks had been instituted, recovering less than one gigalitre (GL)[6] of environmental water (Alberta Environment 2010). Compared to the massive volumes of recoveries undertaken in Australia and California, this volume is infinitesimal. Moreover there is near universal stakeholder agreement that the conservation holdback mechanism has been, and will continue to be, insufficient to meet the established water conservation objectives.[7]

Overall, despite more than two decades of pressure from environmentalists and a major overhaul of the province's water legislation, little environmental water has been recovered in southern Alberta. Conservation holdbacks have provided a token amount of water recovery, but little more than that. The irrigators have been able to protect their water allocations because they are advantageously positioned as an electorally and economically important constituency within the province, and water allocation has been treated as an almost exclusively provincial issue. The environmentalists, though they have become recognized as legitimate stakeholders in water allocation issues, have been unable to match the political influence of the irrigators at the provincial level and have not had the option of pursuing their cause at the federal level. This has put the environmentalists in Alberta at a political disadvantage, resulting in little environmental water recovery.

ENVIRONMENTAL WATER RECOVERIES IN NEW SOUTH WALES

Since the early 1990s, the Commonwealth government of Australia has become increasingly drawn into water allocation issues in the Murray-Darling Basin, issues which had long been the preserve of the states. Commonwealth involvement first took the form of multilateral, intergovernmental initiatives and eventually evolved into unilateral, national ones. The Commonwealth's increasing involvement was driven by a public perception that the Murray-Darling was an issue of national importance most appropriately dealt with at the federal level and by the desire of a succession of Commonwealth governments to score political points with Australia's urbanized, green-conscious electorate. It was also driven, in part, by environmentalists themselves who believed – correctly, as it turned out – that they would have more success in pursuing environmental water recoveries at the Commonwealth level than at the state level. As the past two decades have shown, the more involved the

Commonwealth has become in water allocation issues, the more success environmentalists have had in securing environmental water recoveries.

Serious Commonwealth involvement in intrastate water allocation issues began in the early 1990s through the negotiation of two intergovernmental water reform agreements: the Murray-Darling Basin Initiative (1990) and the Council of Australian Governments (COAG) Water Reform Framework (1994).

The Murray-Darling Basin Initiative involved the four states sharing the Murray-Darling River Basin (New South Wales, Queensland, Victoria and South Australia), the Australian Capital Territory and the Commonwealth. It created a basin-wide intergovernmental institutional architecture for addressing the Murray-Darling's persistent water scarcity and environmental problems. Under its auspices, a number of landmark intergovernmental agreements were reached, such as the Murray-Darling Cap on Diversions and the Murray-Darling Salinity and Drainage Strategy. Though well intentioned and serving as a means to put political and moral pressure on the states to enact water reforms, the Initiative was continually handicapped by its promissory nature and lack of enforcement powers to get governments to honour their intergovernmental commitments (Connell 2007, 113–31).

The other major intergovernmental water initiative of this time, the COAG Water Reform Framework, did not suffer nearly so much from this problem because the Commonwealth used its considerable spending power to enforce its implementation. The Water Reform Framework was an intergovernmental agreement between all of the Australian states and territories to reform and harmonize their water allocation and regulatory systems so that they were more market and environment friendly. The Commonwealth spearheaded this initiative and strongly incentivized state implementation of the agreement by promising them a series of three substantial 'tranche' payments for their compliance. This financial incentive, most observers agree, played a greater role in stimulating state water reform than the promissory agreements of the Murray-Darling Basin Initiative (Connell 2007, 131–4; Quiggin 2001, 75–6).

In terms of water allocations, the Murray-Darling Cap on Diversions and the COAG Water Reform Framework were important because they initiated intrastate water reform processes that resulted in environmental water recoveries. In New South Wales, the state met its intergovernmental water reform commitments by passing and enacting the Water Management Act (2000). Under this legislation, environmental protection was defined as a legitimate and high priority water use, higher even than the highest priority irrigation entitlements. To find the water necessary for environmental protection, the legislation also called for sub-basin level water planning processes with heavy involvement from local stakeholders. The aim of these processes was to create water sharing plans for each sub-basin in which environmental allocations could be defined and reconciled with the existing allocations of irrigators (Productivity Commission 2003). Particularly in sub-basins that were already over-allocated, the net effect of the water sharing plans was to roll back water allocations to the lowest priority irrigators and reallocate this water to environmental protection.[8] Irrigators resisted the rollbacks vigorously but, facing a state government determined to secure its tranche payments from the Commonwealth, were largely unsuccessful (Heinmiller 2007).

Experience with the Water Reform Framework convinced environmentalists that to recover environmental water on a large scale they needed both Commonwealth authority and Commonwealth money on their side to cajole reluctant states, which were generally quite sympathetic to the irrigators. Their first success in large-scale water reallocation came on the Snowy River in the early 2000s. After more than a decade of lobbying by environmentalists and Snowy residents, the Commonwealth, New South Wales and Victorian governments agreed to return a total of 212 GL of water to the Snowy by the year 2012. This water was to be recovered mostly through government-funded improvements to irrigation infrastructure that would allow irrigators to use water more efficiently and reallocate the resulting water savings to the environment. All three governments put up money for the Snowy water recovery scheme, but the Commonwealth paid the largest share (New South Wales Office of Water 2010). The environmentalists' other major success of this period was the Living Murray First Step, and it also involved a substantial Commonwealth leadership role. Negotiated between 2001 and 2004, the Living Murray First Step was an intergovernmental agreement to return 500 GL of water to the River Murray of which 249 GL was to be recovered in New South Wales. The water was to be recovered through a combination of infrastructure improvements and water entitlement buybacks, mostly funded by the Commonwealth government (Lee and Ancev 2009).

With each new initiative the Commonwealth seemed to become increasingly assertive in water allocation issues and it was largely at its insistence that a successor to the COAG Water Reform Framework, an agreement known as the National Water Initiative, was negotiated in 2004. The Commonwealth's assertiveness is largely explained by the onset of a severe drought in the Murray-Darling Basin in the early 2000s, a drought that exacerbated ongoing environmental problems and made them high-profile, national issues. With an election approaching in October 2004, the Howard government was seeking ways to gain support among Australia's predominantly urban and green-conscious electorate and water was the perfect issue (Brett 2011, 35–41). At the Commonwealth's behest the National Water Initiative continued and expanded many of the provisions of the earlier Water Reform Framework, including efforts to recover environmental water in overallocated sub-basins. Notably the Commonwealth also increased its presence in water regulation by creating a National Water Commission to audit state compliance with the new agreement and by committing $2 billion in support of water reform, most of this money dedicated to recovering environmental water through infrastructural improvements (Intergovernmental Agreement on a National Water Initiative 2004; Connell 2007, 151–6). By mid-2005 many stakeholders began to wonder aloud if the Commonwealth was laying the groundwork for a takeover of water regulation in the Murray-Darling. Their suspicions proved to be well founded.

The Commonwealth move came in early 2007 with the announcement of its National Plan for Water Security. Faced with sagging poll numbers, an election later in the year and a continuing drought that had reached crisis proportions, the Howard government made a bold move by seeking to 'reset' water allocation in the Murray-Darling Basin on a sustainable footing.[10] An unprecedented $10 billion of Commonwealth money was committed to support the National Plan for Water Security, which promised to recover environmental water through a massive program of infrastructure improvements and a somewhat more modest program of water entitlement buybacks. The legislative centrepiece of the Plan was the Commonwealth Water Act, which passed through the federal parliament in August 2007 and empowered the federal government to create a Murray-Darling Basin Plan, which would define sustainable water allocation levels for the basin and compel the states to comply with them by 2011 (Government of Australia 2007). While the Commonwealth eventually secured state agreement to this substantial intervention in Murray-Darling water allocation, it had been prepared to proceed unilaterally and actually did so for over a year before securing state agreement in mid-2008 (Crase 2011, 86).

Although Howard's Liberals lost the 2007 election and were succeeded by two Labor governments, the Commonwealth commitment to recovering environmental water in the Murray-Darling has remained firm. In 2008 the Rudd government committed over $2 billion more to the original plan and relaunched it under the banner of Water for the Future (Quiggin 2010). Under both the Rudd and Gillard governments, the Commonwealth has proceeded with studies and public consultations in preparation for the much anticipated Murray-Darling Basin Plan. An early draft released in October 2010, called for 3000 to 4000 GL of water to be reallocated from irrigation to the environment in order to put the basin on a sustainable footing. The scale of such water recoveries is staggering and provoked a strong reaction from irrigators who protested and burned copies of the guide to the proposed plan at public consultation meetings (Crase 2011). Political fallout from the backlash sent Commonwealth administrators back to the drawing board and, at the time of writing, a new draft Plan had yet to be released.

Clearly, over the past two decades a very substantial amount of environmental water has been recovered in New South Wales (and in the other states of the Murray-Darling). The precise amount of water recovered is difficult to say since figures are not available from the early water sharing plans and most of the subsequent, larger reallocation programs are still ongoing. Nevertheless the scale of water recoveries in New South Wales is at least in the hundreds – if not thousands – of gigalitres, a scale that far eclipses the less than one gigalitre of water recovered in southern Alberta. Driving this large-scale reallocation has been an increasingly influential environmentalist movement targeting much of its efforts at the Commonwealth government. Time and again Commonwealth leadership and money were instrumental in prodding the states to adopt and implement environmental water recoveries and, as Commonwealth involvement in water allocation issues steadily increased, so did the amounts of water recovered for the environment.

ENVIRONMENTAL WATER RECOVERIES IN CALIFORNIA

In California, federal involvement in intrastate water allocation issues has a longer history than in either Alberta or New South Wales. The main reason for this was the federal power to preempt state water policies, which, although used sparingly prior to the 1970s, provided a strong constitutional basis for federal intervention in intra-state water allocation issues, something that did not exist in Canada or Australia (Getches 2001). When the environmentalist movement emerged as a political force in the US in the late 1960s, it recognized the potential of the federal preemption power and used it to great advantage. The preemption power allowed environmentalists to make end runs around state policy processes, where their efforts to achieve environmental water recoveries were largely frustrated. The prospect of preemption also served as a powerful incentive to bring states into negotiations to recover environmental water where they would not have done so on their own. In short, the federal presence in water allocation issues has worked much to the advantage of environmentalists and most of the environmental water recoveries in California have stemmed, in one way or another, from Washington's involvement.

California water politics is dominated by three main sets of interests: irrigators; environmentalists; and cities. At the state level, urban and irrigation interests are typically the most politically powerful. Rural areas are dominated by irrigation-based agriculture and perennially elect irrigation friendly representatives to the state legislature. Urban areas, particularly in the south, are constantly preoccupied with securing water resources to serve growing populations and have been quite aggressive in doing so. The most powerful single player in California water politics is the Metropolitan Water District in the greater Los Angeles area, which provides water for around 20 million people and has been quite ruthless in securing its water supplies. Though the irrigators and cities periodically come into conflict over the allocation of specific water resources, they both rely on the extensive infrastructure of dams, canals and pumps that moves vast amounts of water from the Sacramento and San Joaquin rivers in the north of the state to the more arid south. This has often put them collectively at odds with environmentalists and left environmentalists facing formidable opposition in state water policy processes (Ingram and Fraser 2006, 92–4). Consequently, few environmental water recoveries in California have been achieved at the state level.

Irrigators and cities also dominated water policy processes at the federal level until the 1970s when environmentalists managed to break their stranglehold. For decades federal water policy was monopolized by an iron triangle consisting of the water industry (irrigators and cities), western congressmen and senior bureaucrats from the Bureau of Reclamation and the US Army Corps of Engineers. The iron triangle used its monopoly to embark on an unprecedented program of dam building and irrigation expansion, all funded with public money. However by the early 1970s, these policies came under intense attack from fiscal conservatives critical of the large public expenditures, from easterners tired of subsidizing western agriculture and, most of all, environmentalists (Gottlieb 1988, 73; Frisch and Kelly 2008). The old iron triangle lost its monopoly on federal water policy and the federal presence in western water governance was gradually transformed from one of dam subsidization and construction to one of environmental regulation. By the end of the decade, federal funding for new dam projects had all but dried up and new environmental legislation such as the Wild and Scenic Rivers Act (1968), the Clean Water Act (1972) and the Endangered Species Act (1973) had been passed, signalling this change (Getches 2001, 17–18).

On many western rivers, this spate of federal environmental legislation has effectively resulted in environmental water recoveries, typically at the expense of irrigators holding low priority water rights. The Wild and Scenic Rivers Act, the Clean Water Act and the Endangered Species Act empower the federal government to impose water diversion limits on most major rivers for a variety of environmental protection purposes: to preserve river flows; to protect water source quality and to preserve the aquatic habitats of endangered species. When such limits come into effect, the amount of water available to water rights holders is reduced and those holding the lowest priority water rights – typically irrigators – find their allocations cut (Getches 2001).

To avoid such heavy-handed federal preemption, the California government has, over the last two decades, cooperated with federal agencies in order to recover environmental water in a more orderly and voluntary fashion. One of the factors prompting this cooperation was the listing of the delta smelt – a fish species that lives only in the delta formed by the Sacramento and San Joaquin rivers – as an endangered species under the federal Endangered Species Act in the early 1990s. In 1999 it was publicly revealed that the delta smelt population was under threat from the pumping plants that lift water from the Delta and into the canals that service the irrigators and cities of southern California. Federal intervention to protect the delta smelt was mandated, but with the prospect of severe water allocation cuts to southern cities and irrigators, something that both levels of government wished to avoid (Ingram and Fraser 2006, 86–7).

To manage this situation and avoid its recurrence, 25 state and federal agencies came together in 2000 to form the CALFED initiative, which included a new water recovery program known as the Environmental Water Account. The overall objectives of CALFED were to enhance the reliability of water supplies, protect drinking water quality, improve the ecological health of the Delta and strengthen the water infrastructure system (California Bay-Delta Program 2000). The Environmental Water Account contributed to these objectives by recovering water to protect endangered fish stocks through the purchase of water rights from willing sellers (Hanemanna and Dyckman 2009). By one estimate, about 37 GL of water had been purchased for the Environmental Water Account by 2008–9, mostly with federal money (Garrick et al. 2011, 172). Though innovative and fairly widely supported by stakeholders, the program has been undermined by a lack of stable and substantial funding, a problem that has gotten worse in recent years as the fiscal health of both levels of government has deteriorated (Hanemanna and Dyckman 2009).

Even without substantial funding, the federal presence has loomed large in environmental water recoveries. In 2007, as the CALFED process seemed to stall and the delta smelt remained under threat, environmentalists brought suit under the Endangered Species Act. A federal court judge ruled that plants pumping water from the Delta must reduce their water diversions by a third to limit the number of smelt being killed. Already facing water shortages due to drought, the ruling brought swift and protracted protest from irrigators in the San Joaquin valley, who were hardest hit. Nevertheless officials from the federal Fish and Wildlife Service confirmed the scientific soundness of the judicial decision in 2008 and a substantial environmental water recovery for the Delta was effectively and forcibly implemented (Of Farms, Folks and Fish 2009).

Federal action to achieve environmental water recoveries in California has also extended beyond the early federal water laws, as evidenced in the 1992 Central Valley Project Improvement Act. The Central Valley Project is a huge federally constructed irrigation project dating back to the 1930s. It relies largely on water diverted from the Sacramento and San Joaquin rivers through the massive north-south diversion and canal system described above. In the late 1980s environmentalists began lobbying Congress for a significant environmental water recovery from the Central Valley Project in order to restore natural flows and revive native salmon populations in the rivers servicing the project. Despite vocal opposition from several agricultural groups, power generator groups and even California governor Pete Wilson, the Central Valley Project Improvement Act was passed by Congress and signed into law by President George H.W. Bush in late 1992 (Dunning 1993). It requires that water allocations to national wildlife refuges and state wildlife management areas be given a top priority in project water allocations and that approximately 986 GL (800,000 acre-feet) of project water be set aside each year for habitat restoration in the Sacramento-San Joaquin river system (Gillilan and Brown 1997, 231; Garrick et al. 2011, 172). This is the largest single environmental water recovery in any of the jurisdictions analyzed here.

Overall, a very substantial amount of environmental water has been recovered thus far in California and it is pretty clear that the federal government has played a major role in these recoveries. Facing almost insurmountable political opposition from irrigators and cities at the state level, environmentalists have turned to the federal level to achieve their water recovery goals. Sometimes they have used the federal government to make end runs around state processes, while at other times they have used federal pressure to prompt state action. Either way, the federal presence has been an essential factor in facilitating water recovery efforts in California, the nation's largest irrigation economy.

CONCLUSION

Since the emergence of the environmentalist movement in the late 1960s, environmental water recoveries have become a major political issue in irrigation economies such as Alberta, New South Wales and California. Although environmentalists have pressured for water recoveries in all of these jurisdictions, the amount of water actually recovered has varied considerably. The main distinction in this regard, is between Alberta, on the one hand, and New South Wales and California, on the other. In Alberta, less than one gigalitre of water has been recovered from irrigators for environmental protection purposes and the recovery mechanism in place offers little prospect for large water recoveries in the near future. In New South Wales and California, by contrast, hundreds of gigalitres of water have been recovered thus far and programs are in place to recover even more water in the immediate future. Part of the discrepancy between these two groups of cases might be explained by the sheer scale of the rivers involved or the extent of their overallocation problems: one would expect more water to be recovered in larger or more overallocated rivers. However, the rivers of New South Wales and California are not hundreds of times larger or hundreds of times more overallocated than the rivers of southern Alberta: scale alone cannot explain the discrepancies here. Instead environmentalists have simply had much more success in securing environmental water recoveries in New South Wales and California than in Alberta.

This chapter has argued that a crucial factor in explaining the environmentalists' widely varying success in securing environmental water recoveries has been the level of federal involvement in water allocation issues. Federal involvement seems to be positively correlated with environmental water recoveries, for a couple of reasons.

The first reason relates to venue shopping. In irrigation economies, irrigators tend to have a lot of economic and political clout at the state/province level, so much so that they can block most of the environmentalists' water recovery efforts at this level. This was clearly apparent in the cases recounted here as there was not a single large-scale water recovery program undertaken by the Alberta, New South Wales or California governments on their own initiative. Federal involvement opens up a new policy venue for environmentalists, one where local irrigators do not have nearly as much influence as they enjoy at the state/provincial level, giving them a fighting chance to achieve their water recovery goals.

The second reason has to do with the greater spending powers that typify federal governments. Even if the state/provincial governments were amenable to large scale water recoveries, environmentalists might still prefer to pursue their water recovery objectives through federal governments simply because they have more spending power than their state/provincial counterparts. Implementing environmental water recoveries through such measures as water infrastructure improvements or water entitlement buybacks can be a very expensive proposition, so expensive that it may be beyond the financial reach of state/provincial governments alone. Hence the very important role that Commonwealth money played in initiating and sustaining water recoveries in New South Wales and the role that federal money played in initiating the Environmental Water Account in California. Just as federal money was instrumental in expanding irrigators' water use, so it also is in reducing it.

These findings contribute to a small but growing literature, mostly on American federalism, that links federal involvement in resource management issues with positive environmental outcomes. This literature challenges some of the accepted wisdom of subsidiarity, that resources are best managed at lower levels of governance with heavy involvement of local resource users. Local resource users, such as irrigators, can certainly contribute positively to resource management, but they may also capture lower level governance processes to serve their own self interest. In this situation, those wishing to challenge local resource users in the name of a broader public interest – such as environmentalists – stand little chance in lower level processes and need access to higher level processes to pursue their cause. This finding should be kept in mind as political scientists continue to build their theories of multilevel environmental governance.

NOTES

1. 'Full development' aimed to put all available water into productive economic use rather than let it be 'wasted' in the environment; this idea was pervasive in all three countries. 'Closer settlement' was an Australian term describing the settlement of the sparsely inhabited interior of the Australian continent through the creation of small, family-based irrigation farms and rural communities. This idea also shaped water policy in the American West for a number of decades.
2. Federal spending outside its legislative jurisdiction is a longstanding practice in all three federations, and the constitutionality of this practice has been confirmed by the courts. In Canada some relevant cases are *Reference re Canada Assistance Plan* (1991) and *Lovelace v. Ontario* (2000). In Australia the key decision was in the *Second Uniform Tax* case of 1957; and, in the United States, this precedent was set in the *Butler* case of 1936.

3. The riparian rights doctrine states that owners of land that abuts or overlies a water source (i.e. riparians) are entitled to 'reasonable use' of that water source. Since the provinces own most of the public land within their respective borders, this also makes them major water rights holders.
4. Prior appropriation is a water rights system defined by 'first in time, first in right.' Anyone who accesses water and puts it to 'beneficial use' can claim a water right. In times of shortage, water allocations are based on seniority, with the most senior rights (those claimed earliest) receiving full allocations before junior rights receive any allocations.
5. Alberta uses a prior allocation approach that prioritizes water entitlements according to the date they were licensed, with the oldest entitlements prioritized first. So if environmental protection were prioritized ahead of irrigation, less water would be available to meet irrigators' entitlements, and the more recent (low priority) entitlements would bear the brunt of this reduction.
6. A gigalitre (GL) is one billion litres. All water recoveries have been expressed in gigalitres to facilitate cross-case comparison. To put this volume into some perspective, one gigalitre is roughly equivalent to the carrying capacity of three modern supertanker ships.
7. This assertion is based on interviews with irrigators, environmentalists and Alberta government officials conducted during July 2010.
8. At this time, New South Wales had three categories of water entitlement, ranging from highest to lowest priority: high security entitlements; general security entitlements; and off-allocations. Those irrigators relying on off-allocations and general security entitlements have borne the brunt of environmental water recoveries.
9. The Snowy is not naturally part of the Murray-Darling Basin, but was diverted to the River Murray to support irrigation expansion in New South Wales and Victoria through a massive damming project undertaken in the 1950s and 1960s.
10. These assertions are based on stakeholder interviews conducted in July/August 2011.

REFERENCES

Alberta Environment (2010), *Southern Region Completed Transfers Summary.*

Bankes, N. and A. Kwasniak (2005), 'The St. Mary's Irrigation District Licence Amendment Decision: Irrigation Districts as a Law unto Themselves,' *Journal of Environmental Law and Practice*, 1–14.

Brett, J. (2011), 'Fair Share – Country and City in Australia,' *Quarterly Essay* 42, 1–68.

Brownsey, K. (2008), 'Enough for Everyone: Policy Fragmentation and Water Institutions in Alberta,' in M. Sproule-Jones, C. Johns, and B. T. Heinmiller (eds), *Canadian Water Politics: Conflicts and Institutions*, Montreal and Kingston: McGill-Queen's University Press, pp. 133–55.

California Bay-Delta Authority (2003), *California Bay-Delta Program Environmental Water Account Multi-Year Program Plan (Years 4-7).* Retrieved from http://www.calwater.ca.gove/content/Documents/library/ProgramPlans/2008/Draft_EWA_Program_Plan_Year_9.pdf.

California Bay-Delta Program (2000), *Programmatic Record of Decision.* Retrieved August 11, 2010, from http://www.calwater.ca.gov/content/Documents/10_Year_Action_Plan_Final.pdf.

Clark, S. D. (1983), 'The River Murray Waters Agreement: Peace in Our Time?' *Adelaide Law Review*, 108–41.
Connell, D. (2007), *Water Politics in the Murray-Darling Basin,* Annandale: The Federation Press.
Crase, L. (2011), 'The Fallout to the Guide to the Proposed Basin Plan,' *Australian Journal of Public Administration* 70 (1), 84–93.
Davidson, B. (1969), *Australia Wet or Dry? The Physical and Economic Limits to the Expansion of Irrigation,* Melbourne: Melbourne University Press.
Dunning, H. (1993), 'Confronting the Environmental Legacy of Irrigated Agriculture in the West: The Case of the Central Valley Project,' *Environmental Law* 23(3), 943–69.
Figliuzzi, S. (2002), 'South Saskatchewan River Sub-Basin Contributions to International and Interprovincial Water-Sharing Agreements', published by Alberta Environment as part of South Saskatchewan River Basin Water Management Plan, Phase Two: 'Background Studies'.
Frisch, S. A., and S. Q. Kelly (2008), *Jimmy Carter and the Water Waters – Presidential Influence and the Politics of Pork,* Cambria: Amherst Press.
Garrick, D., C. Lane-Miller, and A. L. McCoy (2011), 'Institutional Innovations to Govern Environmental Water in the Western United States: Lessons for Australia's Murray-Darling Basin,' *Economic Papers* 30 (2), 167–84.
Getches, D. M. (2001), 'The Metamorphosis of Western Water Policy: Have Federal Laws and Local Decisions Eclipsed the States' Role?' *Stanford Environmental Law Journal*, 3–72.
Gillilan, D. M., and T. C. Brown (1997), *Instream Flow Protection – Seeking a Balance in Western Water Use,* Washington DC: Island Press.
Glenn, J. (1999), *Once Upon an Oldman: Special Interest Politics and the Oldman River Dam,* Vancouver: UBC Press.
Gottlieb, R. (1988), *A Life of Its Own: The Politics and Power of Water,* New York: Harcourt Brace Jovanovich Publishers.
Government of Australia (2007), 'A National Plan for Water Security,'. retrieved 20 June 2011, from http://www.nalwt.gov.au/files/national_plan_for_water_security.pdf.
Hanak, E. (2003), *Who Should Be Allowed to Sell Water in California? Third-Party Issues and the Water Market,* San Francisco: Public Policy Institute of California.
Hanemanna, M., and C. Dyckman (2009), 'The San Francisco Bay-Delta: A Failure of Decision-Making Capacity,' *Environmental Science & Policy* 12 (6), 710–25.
Heinmiller, B. T. (2007), 'The Politics of 'Cap and Trade' Policies,' *Natural Resources Journal* 47 (2), 445–67.
Ingram, H., and L. Fraser (2006), 'Path Dependency and Adroit Innovation: The Case of California Water,' in R. Repetto (ed.), *Punctuated Equilibrium and the Dynamics of US Environmental Policy,* New Haven: Yale University Press: pp. 78–109.
Intergovernmental Agreement on a National Water Initiative (2004), retrieved 5 July 2011, from http://www.nwc.gov.au/resources/documents/Intergovernmental-Agreement-on-a-national-water-initiative2.pdf.

Jaensch, D. (1997), *The Politics of Australia,* South Yarra: Macmillan Publishers Australia Pty Ltd.

Kennett, S. (1991), *Managing Interjurisdictional Waters in Canada: A Constitutional Analysis,* Calgary: Canadian Institute of Resources Law.

Koontz, T. M. (2002), *Federalism in the Forest: National Versus State Natural Resource Policy,* Washington DC: Georgetown University Press.

Layzer, J. A. (2008), *Natural Experiments – Ecosystem-Based Management and the Environment,* Boston: Massachusetts Institute of Technology.

Lee, L. Y.-T., and T. Ancev (2009), 'Two Decades of Murray-Darling Water Management: A River of Funding a Trickle of Achievement,' *Agenda* 16 (1), 5–23.

Lloyd, C. (1988), *Either Drought or Plenty: Water Development and Management in New South Wales,* Sydney: Department of Water Resources, Government of New South Wales.

Mertha, A. C., and W. R. Lowry (2006), 'Unbuilt Dams: Seminal Events and Policy Change in China, Australia, and the United States,' *Comparative Politics* 39(1), 1–20.

Murray-Darling Basin Ministerial Council (1995), *An Audit of Water Use in the Murray-Darling Basin.*

New South Wales Office of Water (2010), *Returning Environmental Flows to the Snowy River – An Overview of Water Recovery, Management and Delivery of Increased Flows*. Retrieved July 7, 2011, from http://www.water.nsw.gov.au/Water-management/Law-and-policy/managing-the-snowy/default.aspx.

'Of Farms, Folks and Fish' (2009), *The Economist*, 24 October, pp. 27–30.

Percy, D. R. (1996–97), 'Seventy-Five Years of Alberta Water Law: Maturity, Demise and Rebirth,' *Alberta Law Review*, 221–41.

Percy, D. R. (2005), 'Responding to Water Scarcity in Western Canada,' *Texas Law Review*, 2091–107.

Productivity Commission (2003), *Water Rights Arrangements in Australia and Overseas: Annex B, NSW,* Melbourne: Productivity Commission.

Quiggin, J. (2001), 'Environmental Economics and the Murray-Darling River System,' *The Australian Journal of Agricultural and Resource Economics* 45(1), 67–94.

Quiggin, J. (2010), *Water Policy After the Drought,* the Risk and Sustainable Management Group. Retrieved 7 June 2011 from http://www.uq.edu.au/rsmg/working-papers-rsmg.

Randall, A. (1981), 'Property Entitlements and Pricing Policies for a Maturing Water Economy,' *Australian Journal of Agricultural Economics*, 195–220.

Reisner, M. (1993), *Cadillac Desert – The West and Its Disappearing Water,* Vancouver: Douglas & McIntyre.

Smith, P. J. (1996), 'The Politics of Alberta's Water Policy Review: Has a New Era Dawned?' Annual Meeting of the Canadian Political Science Association, St Catharines, Ontario.

Worster, D. (1985), *Rivers of Empire – Water, Aridity and the Growth of the American West,* New York: Oxford University Press.

5. Playing a zero sum game: sharing water between jurisdictions in federations

Inger Weibust

Climate change has been raising concerns about water shortages and interstate water wars, even though the most recent case of armed conflict over water occurred 4500 years ago. More recently, in 1934, armed troops massed along a border in a water supply conflict. The border in question was not an international boundary but the one between Arizona and California. Arizona's governor called out the National Guard to prevent California's construction of the Parker dam, which was expected to reduce Arizona's share of the Colorado River (Gleick 2000). The United States (US) Supreme Court eventually settled the dispute (Gleick 2009a).

Although *armed* conflict over water is very rare, other forms of conflict are commonplace when river waters are allocated between jurisdictions directly upstream/ downstream of one another. The game theory term for this zero sum situation is 'pure conflict,' because there is no win/win solution. A unit of water consumed by one jurisdiction cannot also be consumed by another.

Furthermore, river flow allocation can illustrate additional meanings of 'conflict': contention and rivalry. A legal scholar of US interstate water disputes compares these to kickboxing: not lethal but fairly vicious nonetheless (Ruhl 2003, 57). Another legal scholar contends that interstate water conflicts are never 'resolved with finality. Such conflicts have a great deal in common with volcanoes: sometimes they are active, sometimes they are dormant but they are always volcanoes.' Thus, in addition to the 1934 brush with armed conflict, between 1922 and 1966, conflict between US states over the lower basin of the Colorado River resulted in five cases being brought before the United States Supreme Court (Sherk 2005, 815).

Lest anyone think such cases are ancient history, more recent times witnessed a serious and long-running dispute between Georgia, Alabama and Florida. In 2007, a severe drought put Atlanta and other cities in the US southeast within 90 days of running out of municipal water supplies (Goodman 2007). In 2009, a US federal court sided with the downstream states, ruling that Atlanta had been drawing illegally on a reservoir for its water supply. The judge gave Georgia three years to fix its water supply or obtain Congressional authorization. In response, the Governor asserted that Georgia was entitled to use all the river waters originating in Georgia (the legal principle known as the Harmon Doctrine). However water expert Peter Gleick (2009b) observed, '[t]he only problem for Governor Purdue, and Georgia, is that the Harmon Doctrine has been universally repudiated: in international law, in US law and in any ethical or moral set of rules.'

Global climate change may make interjurisdictional water disputes more common and more severe. The upstream/downstream allocation dynamic for rivers is a zero sum game and thus poses a real challenge for cooperation.[1] These situations also aggravate policy makers and politicians for the same reason. This chapter compares the regimes for allocating transboundary river waters in federal or quasi-federal systems: the US; the European Union (EU); and Canada. Each regime is assessed according to accepted criteria for optimal water governance:

1. Compatibility with Integrated Water Resource Management
2. Fairness and equity
3. Certainty and security of resource allocation
4. Dispute resolution
5. Quality of governance.

Each of these federations has regions already under water stress. I will judge the potential of each regime for dividing a shrinking pie among several jurisdictions and coping with more acute contention over supply.

INTERSTATE RIVER ALLOCATION IN THE UNITED STATES

The United States is no stranger to water allocation conflict, particularly in the West. Water law and jurisprudence on interstate river allocation are highly developed there. The US case offers an interesting interplay of cooperation against a backdrop of a higher level of government authority. The US has cooperative interstate agreements on rivers analogous to river treaties at the international level. However, unlike the international situation, between US states there is an alternative to reaching a negotiated agreement. At the international level the alternative is that the upstream jurisdiction does as it pleases:

> Water rights that are specified in [an international] water allocation agreement cannot be enforced by a downstream country. Especially so, because there is no higher authority that can enforce compliance. (Ansink and Ruijs 2008, 252)

This is not so for upstream US states because of US federal law.

The western region of the US, where water has always been scarce, follows the water law rule of 'First in Time, First in Right', known as FITFIR. (FITFIR does not apply in the eastern US.) FITFIR ensures security of access for existing users while disadvantaging more recent users. The rule also heightens conflict between rural and urban users because the oldest uses are almost always withdrawals for irrigation.

Between states and within states such as California, growing cities vie with irrigators to divvy up shrinking water resources. On the Colorado River, one of the most contested interstate rivers, 90 per cent of withdrawals are for irrigation even though all of the following major cities draw at least some of their municipal supply from this river: Los Angeles; Las Vegas; San Diego and Phoenix; San Bernardino and Tucson.

United States law provides four means to resolve interstate disputes about river allocation:

1. Allocation by the US Congress
2. Interstate compacts approved by the US Congress
3. Referral to the US Supreme Court to resolve the dispute
4. Litigation under federal laws such as the Endangered Species Act. (Ruhl 2005, 47)

Allocation by US Congress

Congress has acted only twice to allocate interstate water, far less often than has the Supreme Court (Sherk 2005, 814). This suggests Congress is unwilling to become involved in these matters, perhaps because they pit state against state.

Interstate Compacts for Water Allocation

Interstate compacts govern most allocations of interstate rivers, with at least 14 interstate water allocation compacts in effect in the US (US Interstate Freshwater Compacts Database).[2] These compacts generally protect existing rights and uses, in keeping with FITFIR (Muys 2003). Legally, the provisions of an interstate compact supersede the laws of any individual state (Schlager and Heikkila 2009, 368). Interstate compacts are analogous to international treaties, requiring the consent of all parties but also the consent of the US Congress to enter into effect. Unlike international treaties, which need to be self-enforcing, interstate compacts are enforceable in the US federal courts.

US States began to negotiate river compacts because they were not satisfied with the outcome when courts settled interstate water disputes. Furthermore, the federal Reclamation Act 1902 made the construction of large-scale water projects conditional upon the resolution of interstate water conflicts (Schlager and Heikkila 2009, 369). An additional incentive was states' interest in limiting the powers and jurisdiction of the US federal government, although these powers were limited prior to 1970.

Despite the desire to limit federal involvement, the first water allocation compact (1922) was only made possible after US federal legislation voided the usual unanimity rule. The interstate compact on the Colorado River only came about once Arizona could be overridden, after Congress set a 6/7 voting rule for passage of the compact. Arizona continued to dispute its allocation. The US Supreme Court settled the issue with the 1963 decision, *Arizona v. California*, upholding the original water allocation.

Because of their enforceability, American interstate compacts often give rise to litigation (as per the dormant volcano analogy). Interstate compacts are legal contracts (Muys 2003). Thus, if an upstream state fails to deliver the volume of water promised in the compact, it can be taken to court and forced to pay compensation. In 1990 the US Supreme Court ordered New Mexico to pay Texas $14 million for failing to deliver the volume of water promised in the Pecos River Compact (1948) between the years 1952 and 1986 (Hanada and Hancock 2004, 160).

Because water compacts usually set out precise formulas for allocation, they have the virtue of being specific, providing secure property rights to states. The downside is rigidity, which leads to problems. First, the formulas in many interstate compacts were based on unreliable estimates of supply, often flows in uncharacteristically wet, not typical, years (Muys 2003). Furthermore, for many Western water basins, climate change is leading to reduced flows. Lastly rigidity in allocation makes urban population growth particularly challenging: states want sufficient municipal supplies for urban growth and economic development.

Another major disadvantage of compacts is the lengthy negotiating process, especially if the compact includes an allocation formula. Compacts negotiated in the early twentieth century took many years but so did a more recent attempt. Georgia, Alabama and Florida began negotiations over the Apalachicola-Chattahoochee-Flint river system (ACF) in 1983 (Ruhl 2005, 50). In 1997 Congress authorized a federal-interstate compact between Georgia, Alabama and Florida with the US federal government as a partner. Under the compact, these states, with federal assistance, were supposed to arrive at an allocation formula. After two decades of negotiation and mediation and many deadline extensions, the federal-interstate ACF compact was allowed to expire in 2003, without ever reaching agreement on an allocation formula. In 2011 a federal Appeals Court overturned the ruling that required Georgia to find alternate municipal water supply. In 2012 the US Supreme Court indicated it would not revisit the case. Georgia may continue to use the reservoir to supply Atlanta but the three states are supposed to arrive at an agreement for managing the watershed ('Top-Court' 2012).

Referral to the United States Supreme Court to Resolve the Dispute

No water compacts provide for binding arbitration, thus the US Supreme Court is the only venue for authoritative dispute resolution (Muys 2003). The first such case was *Kansas v. Colorado* (1907), which rejected the Harmon Doctrine that the upstream state is entitled to the entirety of the river's flow (Ruhl 2003, 53). Over time, the doctrine of equitable apportionment replaced it. In 1982 the US Supreme Court ruled that 'the equitable apportionment of appropriated rights should turn on the benefits, harms and competing efficiencies of competing uses, and the source of the … river's waters should be essentially irrelevant to the adjudication of these sovereigns' competing claims' (*Colorado v. New Mexico* (1982)). Equitable apportionment asks the courts to weigh a long list of factors. Most legal commentators agree that equitable apportionment is 'a vague set of standards that are impossible to quantify' (Sherk 2005, 826). It is significant that this principle also forms the basis for international water law on interstate transboundary water allocation.

Litigation under Federal Laws Such as the Endangered Species Act

The 2009 court decision giving Georgia three years to obtain Congressional authorization centred on the federal legislation authorizing a dam to create Atlanta's reservoir. Some commentators have argued that interstate water conflicts can only be resolved by Congress because of the many complex and potentially conflicting requirements imposed by federal laws and regulations applicable to interstate water resources (Sherk 2005, 826).

ASSESSMENT OF THE US REGIME

A complete assessment of the US regime for allocating interstate rivers requires the examination of interstate compacts but also the complex backdrop of legal doctrines developed over a century of Supreme Court decisions and the explosion of post-1970 US federal environment laws pertaining to water allocation.

Integrated Water Resources Management

The regime cannot be said to have been designed with holistic water resources management as a goal. Preeminent compact expert Jerome Muys (2003) indicates that older water allocation compacts 'are so environmentally outdated that is seems unlikely that most of them would currently receive Congressional consent.' The failure to integrate groundwater resources and to consider coastal waters are additional flaws.

There is some room for optimism. The ecological benefits of instream flows have gained recognition (Das Gupta 2008, 299). Older compacts allocated most of a river's flow for consumptive uses. More recently federal legislation such as the Endangered Species Act imposes requirements for water quality with direct implications for river flow and therefore allocation. A solution to the dispute over the ACF river basin is likely to explicitly recognize the value of ecological flows required to maintain marine life in the Apalachicola estuary (Ruhl 2003, 53).

Fairness and Equity

The American system provides robust guarantees for downstream users. Federal law reflects the principle of 'equitable utilization' ensuring that downstream states are not at the mercy of upstream riparians. Recourse to the federal courts ensures that both federal law as well as interstate compacts are enforceable. Where upstream states fail to uphold their obligations, downstream states can be awarded compensation for the damages, in water deliveries or cash, including past damages. This better protects downstream riparians than less formal, more cooperative regimes.

Certainty and Security in Allocations

Interstate compacts offer a high degree of certainty in allocation. In the US West, the oldest uses are protected. However the amount of water actually received will depend on the river's flow if the allocation is proportional to total flow, rather than being a fixed quantity. If however, a dispute over allocation (within or without a compact) goes to litigation, outcomes are less certain.

One analysis of the failed ACF compact identified uncertainty as an impediment to negotiations, including uncertainty about the outcomes of ongoing litigation in the case:

> Increasing legal certainty may thus be helpful for averting future water allocation disputes ... This is clearly true with respect to the Supreme Court's Equitable Allocation doctrine. States deciding whether to pursue water litigation in the Court could benefit from a doctrine that provides more guidance to parties with respect to both the conditions required for the Court to take jurisdiction (e.g., the standard of harm that downstream states must show), and the substantive outcomes favored by the court. (Snowden 2005, 193)

However, another legal scholar argued that uncertainty about litigation was a positive incentive for states to conclude a compact (Grant 2003, 178). These two scholars do agree that the equitable allocation principle encompasses so many considerations that it is very difficult to predict how courts will rule in any given case.

Recourse to Dispute Resolution

Clearly the US system does not lack for dispute settlement although the quality of the decisions may vary depending on the forum. Some compacts have internal procedures. There is also the possibility of mediation by the federal government or third parties as well as enforcement by federal courts. There is scholarly debate over the capacity of consensus-based bodies to manage natural resources and resolve disputes. A study of interstate compact commissions observed that commissions could successfully resolve water allocation disputes, even though these disputes, as zero sum conflicts, are generally considered difficult to address (Schlager and Heikkila 2009, 386). However, the study also found compact commissions inferior to courts in resolving complex conflicts, such as those going beyond simple allocation to include other issues such as water quality.

Quality of Governance

Integrated Water Resource Management makes the assumption that the optimal form of water governance is river basin level management, with authoritative decisions being made by officials and stakeholders at that level. Although interstate compacts govern particular river basins, they do not represent stakeholder centred, river basin focused governance. Decisions are made by state governments at the level of executives and legislatures. States have consistently chosen not to delegate powers to compact commissions (the Delaware Basin Federal Interstate compact is the exception to this rule) (Muys 2003). When compact signatories fail to agree, decisions are imposed, top down, by federal courts. Thus the US regime does not qualify as participatory stakeholder centred governance.

US Summary

The American system of interstate river water allocation has the benefit of over a century of experience. This has given rise to several modalities for managing these perennial conflicts, which give every indication of becoming more common and more severe in the future. The existing regime is effective in delivering authoritative decisions about water allocations. It is also slow, expensive and enormously complicated, given the accretion of decades of water law jurisprudence as well as the multiple demands of federal environmental legislation. This accretion of jurisprudence has helped produce outcomes that better reflect ecological, as well as consumptive values of water resources. Because the system is so tied to legal contracts and their interpretation, it is far removed from the vision of an Integrated Water Resource Management with basin-centred governance and extensive stakeholder participation.

WATER QUANTITY AND ITS GOVERNANCE IN THE EUROPEAN UNION

Water Supply in the European Union

The EU is no stranger to the threat of drought and water scarcity, which climate change is predicted to exacerbate (European Commission 2007). Much of the Mediterranean faces serious water supply pressures even in years of normal precipitation. These regions are squeezed by low rainfall coupled with increased demand due to population growth and tourism, as well as ongoing demand for irrigation. In southern Member States, such as Spain, irrigation may consume as much as 80 per cent of the available water. Regions beyond the arid south also face water stress. These EU Member States were at least partly water stressed in 2007: Belgium; Bulgaria; Cyprus; Germany; Italy; Malta; Spain; and the United Kingdom (European Environment Agency 2009a).

There are 49 international river basins within the EU, out of a total of 128 river basins. Because of the size of international basins such as the Danube, they comprise 60 per cent of the area of all basins. Of the international basins, about 40 lie wholly within the EU (European Commission, DG Environment 2008, 3). There are many river allocation agreements between Member States, some of which may become contentious in the future. (For example, Spain and Portugal share five river basins.)

Thus far, conflicts over river allocations within the EU have been most acute within Member States, such as among regions within Spain. The greatest interjurisdictional supply crisis was Barcelona's severe water shortage in 2008, which saw its reservoirs fall to 20 per cent of capacity due to drought. The Spanish government constructed a temporary pipeline diverting water from the Ebro River (which flows primarily within Spain). In response, the southern provinces of Murcia and Valencia threatened to take the national government to court, in order to force a permanent diversion of the Ebro for their regular water needs.

A permanent Ebro diversion to southern Spain had been rejected previously on economic and environmental grounds. Experts deemed a €6 billion Ebro transfer project 'flawed and extremely expensive' (Garrido and Llamas 2009, 127). The European Commission, which usually subsidizes large infrastructure projects, expressed reluctance because the Ebro transfer would not meet the Water Framework Directive's requirements for cost recovery (Albiac et al 2006). The project to divert the Ebro permanently was scrapped.

Consumptive Uses Within the EU

The EU's major uses of river water are energy production, irrigation and industrial uses. Although cooling power plants is the largest single use, its environmental impact is limited because that water is returned to the watershed. In contrast only 30 per cent of irrigation water returns to the basin (European Environmental Agency 2009, 6). Allocating river water across boundaries may be a lower priority for the EU than the US because most municipal water supplies in Europe rely on groundwater not rivers.

The Cornerstone of EU Water Policy: The 2000 Water Framework Directive

Since the 2000 Water Framework Directive (WFD), EU water policy has mandated river basin level governance, as well as efficient use through water pricing. The WFD set a deadline of 2009 for Member States to adopt and present their River Basin Management Plans (RBMP). The directive also requires public participation in basin governance, primarily through stakeholder consultation.

Some members have experience of river basin level management but for most countries it represents a significant transition. In a survey of WFD implementation, environmental law scholar Elli Louka (2008, 66) found that the WFD simply added another institutional layer into whatever constellation of water governance was already in place in a particular Member State. The WFD has a good fit with those Member States that had already instituted water governance at the river basin level, such as the United Kingdom and France.

As of May 2013 Belgium, Spain, Portugal and Greece had yet to submit all their required RBMPs (European Commission 2013). Belgium and Spain had adopted some but not all their plans; Greece and Portugal had not reported any plans. The European Court of Justice had already ruled against Greece, Portugal and Belgium for their failure to report (European Commission 2012b, 5). These four Member States are all considered water stressed and all contain at least one river shared with another Member State.

In terms of water quantity (rather than quality), the objective of the WFD is to prevent water scarcity through demand management. A European Environment Agency report on confronting water scarcity and drought (2009b, 6) promoted strategies such as water pricing and reducing losses due to leaky pipes and illegal irrigation withdrawals. As of 2009, Member States were required by the WFD to meter all uses of water and charge users by volume, not simply an access fee. The 2012 follow up report on the implementation of the WFD noted little progress in transparent water pricing and found many Member States had yet to meter agricultural water use (European Commission 2012b, 10–11). A 2012 study of Member State policies found that prices for irrigation water did not reflect the full costs (European Commission 2012a, 90).

Significantly, the EU lacks explicit policies on water allocation between jurisdictions, either within or between Member States. The WFD stipulates that all water withdrawals from river basins must be authorized by the relevant government authority (Holwarth 2002). However experience from the US and elsewhere suggests that requiring licences for withdrawals is not sufficient for ensuring adequate supplies, particularly beyond the jurisdiction issuing the licences.

The WFD has been criticized for emphasizing water quality over quantity. The WFD directive encourages, but does not obligate, Member States to develop Drought Management Plans where necessary (Article 13(5)). According to scholar Elli Louka (2008, 66), 'Southern European states have charged that the directive addresses the environmental issues of northern Europe without addressing the water management issues of the south. In the south issues of water quantity are more acute than those of water quality.' The 2012 follow up report on implementation of the WFD contains very few pages on water quantity and scarcity in relation to the extensive reporting on water quality (European Commission 2012b).

Water Allocation Agreements and Conflicts Between Member States

Although river allocation has not been a high profile issue within the EU, there have been conflicts between members over river water allocation. According to the International Freshwater Treaties Database, treaties covering river allocation between EU countries range in date from 1863/1873 (amended 1961) for the Meuse River, to 1995 for the Nestos River flowing from Bulgaria into Greece (Oregon State University 2007). Although some of these rivers pass through more than two countries, allocation treaties between EU members tend to be bilateral treaties.

For decades Belgium and the Netherlands had an allocation conflict over the Meuse, which supplies water to Brussels, Antwerp, Rotterdam and other towns. The conflict was eventually resolved by linking it to other issues (navigation) in another shared river basin (the Scheldt), as well as water quality concerns in both basins (Mostert 2001). Belgium's federal structure raised significant obstacles to the agreement. Draft agreements reached in 1975 were never signed because, while they would produce benefits for the Netherlands and some Belgian regions, the Walloon region faced only increased costs (for reservoirs) and no benefit. Although joint management plans have been created in the last 15 years, cooperation remains elusive.

It might be expected that European integration or expansion would hasten cooperation or integration in water supply governance but this is not necessarily so. There are still international river basins lying entirely within the EU that lack any 'formalized cooperation,' specifically two rivers between Greece and Bulgaria (European Commission, DG Environment 2012). A third basin between these two Member States, the

Mesta/Nestos has a formal agreement but no cooperative body or international RBMP.

The 1995 Mesta/Nestos River agreement between Bulgaria and Greece (signed long before Bulgaria's 2007 accession to the EU) is quite underdeveloped. It allocated 29 per cent of the river's flow to Greece, but without specifying whether the flow is annual or monthly (Mylopoulos, Eleftheriadou, Kampragou 2004). This distinction has significant consequences for the quantity of water to be supplied. Most importantly, the agreement made no provisions for allocation under drought conditions, when allocation is most likely to be a critical issue (UNEP 2009, 39).

Between EU Member States, river water allocation remains a state to state affair. Even where subsequent international agreements have been negotiated for a river, these have not superseded or incorporated decades old bilateral water apportionment agreements. In theory, the EU could have been designated an arbiter for water disputes between Member States, but this has not occurred. Experience with allocation disputes suggests that identifying a dispute resolution body is important for robust agreements and effective dispute resolution.

Water Allocation Governance at the Level of the European Union

The EU has no specific policy on allocating water between riparian Member States nor among regions within Member States. Member States are required to take hydrology and the overall ecology of waters into consideration in governing water. The WFD mandates the creation of basin level river governance but thus far Member States have chosen to vest little authority in these institutions (Louka 2008, 402). As of 2012, only four Member States granted any enforcement powers to their river basin boards; these are primarily states whose river basin governance predated the WFD (European Commission 2012c, 37).

Although the EU WFD provides a framework for water policy in the EU, in the allocation of river flow, the EU neither directs nor seriously constrains national actions, either within or between Member States. The WFD requires that basin wide governing bodies be created for all international rivers (International RBMPs). As of 2012, these International RBMPs have been created for only eight rivers: the Elbe, Ems, Danube, Meuse, Oder, Rhine, Scheldt and Shannon (European Commission, DG Environment 2012).

The WFD specifies that management of transboundary catchments shall be based on existing international agreements, such as the 1992 UN ECE Convention on the Protection and Use of Transboundary Watercourses and International Lakes. Most of these texts are based on the principle of equitable use. However water law jurisprudence is clear that equitable use does *not* mean equal use or an equal allocation between riparian states: 'Equality in use does not mean equal shares but an "equitable share"' (del Castillo-Laborde 2009, 321).

A survey of freshwater conflict and cooperation notes that this language is not helpful in determining allocations between parties:

> Generic international water law usually offers limited guidance because the main principles 'no significant harm' and 'reasonable and equitable utilization' are quite abstract and may conflict with each other in individual cases, especially in highly developed basins where the present utilization could be interpreted as not equitable. (Mostert 2003, 36)

Even the more recent Berlin Rules (2004), offer little concrete guidance, or at least, do not provide firm guidelines that would produce more predictable decisions.

For example, the Berlin Rules state allocations must prioritize 'vital human need.' That provides no guidance in resolving a dispute between irrigation or industrial uses in two jurisdictions. In Europe this type of conflict (or two irrigation uses) is far more likely than one in which municipal water supplies are threatened. Article 13 of the Berlin Rules offers a long, but not exhaustive, list of nine very broad factors to be considered, without any hierarchy in the factors. It seems likely that in many disputes either side could cherry-pick factors advantageous to its case. Thus, referring a dispute to international law offers no definitive solution to authoritative allocation or conflict resolution.

ASSESSMENT OF THE EU'S REGIME

Integrated Water Resources Management

The Water Framework Directive has been praised by many experts for its explicit efforts to promote Integrated Water Resource Management. It institutes governance at the basin level and incorporates coastal and groundwater impacts, not just surface water. It also gives explicit recognition to the value of ecological flows, although much work remains in implementing this principle (European Commission 2012b, 9). The WFD's primary emphasis on demand management and prevention may not be sufficient to avoid scarcity. For example Spanish research on pricing irrigation water found that irrigation price increases made farmers economically worse off but did little to reduce their water use. Where demand for irrigation water is inelastic, a shift to water pricing will not ensure that there is adequate supply to meet demand (Garrido and Llamas 2009, 136).

Despite water experts' enthusiasm for basin level governance, in allocation, the bilateral state-to-state approach may have merit. The Meuse agreement between Belgium and the Netherlands was only made possible by linkage to other issues on their other shared river (the Scheldt). Governance structures that emphasize holistic basin governance more than high politics and deal making might jeopardize such agreements. If river governance is insulated from other policy issues, the possible range of linkages is greatly limited and cash side payments might be the only remaining option. The fundamentally zero sum nature of water allocation will rear its ugly head.

Fairness and Certainty

Because no transboundary allocations have been disputed in recent years, it is difficult to assess the EU's regime in terms of fairness or certainty. (The proposed Ebro transfer did not concern allocation of the natural flow of a river but rather an artificial diversion.) In theory the legal doctrine of equitable apportionment should apply in the resolution of any international dispute but that does not offer predictability in terms of the outcome, nor any certainty of enforcement of any ruling in international law.

Dispute Resolution

In the event of a serious water shortage or major riparian dispute between Member States, it seems likely the EU would be drawn into the dispute, if only as a mediator. Given that drought planning on water basins is only suggested by the WFD, it is possible that an upstream state on a shared river basin does not have a drought plan. Or, even if there is a drought plan, the upstream state may be unable or unwilling to fully implement it in the face of domestic opposition. Elected officials only have to face their own voters, not those downstream in another jurisdiction.

In the long run, the European Court of Justice may be called upon to make authoritative rulings in disputes between Member States. The 1995 Mesta/Nestos agreement between Bulgaria and Greece makes no provision for enforcement or conflict resolution. Similarly there are no provisions for enforcement or conflict resolution in the agreements between Belgium and the Netherlands over the Meuse (Oregon State University 2007). Although international law provides rules for such cases, these rules lack centralized enforcement capacity, unlike the EU.

Experience in the US indicates that water users want authoritative decisions about water allocation, to enable planning. Users want to avoid uncertainty and thus are likely to gravitate to governing bodies that have the power to make authoritative decisions.

It is also noteworthy that both for the Meuse/Scheldt agreement and Mesta/Nestos agreement, subnational governments played a significant role in preventing agreement or implementation of a negotiated agreement. International law is unlikely to be of use for this type of problem: subnational governments do not enjoy legal 'personality' in international law. International law is quite clear that subnational entities are obliged to comply with international agreements but there are few tools in international law for punishing subnational governments for non-compliance.

Summary EU

At present, the EU is not directly involved in allocating water in rivers, even when more than one Member State is involved. That responsibility rests solely with Member State governments. The focus of the WFD is wholly on demand management to prevent water scarcity. In any dispute over international watercourses, the international law principle of equitable apportionment applies, which should ensure a fair outcome. There is no evidence to date that the EU aspires to settle or even mediate future disputes between Member States. For rivers in their natural state, the EU cannot influence member behaviour by withholding funding because, unlike dams or diversions, the river in its natural state does not require financial support from the EU. It is unclear that EU policies alone will be sufficient to prevent contention over water allocation, nor that the EU is equipped to resolve such disputes should they arise.

ALLOCATING WATER BETWEEN CANADIAN PROVINCES

Status of Canadian Water Supply

Although Canada's rivers discharge 9 per cent of the global flow of freshwater, most of it flows north, away from the majority of the Canadian population (Schindler 2001). A few of Canada's major rivers cross a provincial or territorial boundary: the North and South Saskatchewan, the Mackenzie and the Peace. Several regions of the country experience water scarcity and drought, particularly the Prairie provinces (Alberta, Saskatchewan and Manitoba):

> The 20th century was climatologically benign compared to other recent centuries. While significant droughts have taken place, for example in the 1930s, they may not have been as severe as those of previous centuries. Nonetheless, the 20th century featured three major Prairie drought events. (Partners 2009a, 15)

Thus, even without climate change, a reversion to more typical weather patterns of the past lead us to expect reduced water supplies.

Furthermore climate change is predicted to reduce water supplies to the Prairies, if the snow pack in the Rocky Mountains declines (Schindler 2001) (Draper and Kundell 2007, 413). These pressures are likely to be acute for the South Saskatchewan River (SSR), which originates in the Rockies and flows through Alberta and Saskatchewan, eventually joining the Saskatchewan River and flowing into Manitoba.

A survey of SSR basin stated:

> The current water-supply situation prompts questions about water supplies under future climate scenarios and raises concerns about environmental effects of existing and proposed dams and diversions. (Partners 2009a, 11)

At present, during low flow years, Alberta delivers just enough water in the SSR to meet its obligation to Saskatchewan under an interprovincial agreement. A 2003 article assessed the supplies on the North Saskatchewan River as well as the southern reaches of the Saskatchewan River as ample (Bruce et al. 2003, 92). However, the Battle River, an interprovincial tributary of the North Saskatchewan has been identified as one of the most heavily exploited in Canada (Environment Canada 2007).

In a median year about 35 per cent of the South Saskatchewan River's naturalized flow is consumed, first and foremost by irrigation (Partners 2009a, 11). Of all the water consumed in the entire Saskatchewan River basin (North and South), irrigation withdrawals from the SSR are the largest use. The flows of the Bow and Oldman Rivers (sub-basins of the SSR lying within Alberta) are considered fully allocated; some consider them already over-allocated (Saunders and Wenig 2007, 128). Additional withdrawals from these basins would have negative consequences for the rivers. As of 2006, the Alberta government no longer issued new irrigation licences for these two sub-basins but this measure may be insufficient to manage demand.

Alberta's system for water licences, based on FITFIR, has been heavily criticized:

> Alberta granted the bulk of these allocations through water licenses that are decades old, have no expiration dates, and were likely issued with little or no thought to their effect on watershed ecosystem health. Notwithstanding the antiquated management philosophies underlying these licenses, to date, Alberta has generally refused to consider taking back those historic licenses (or, at least, the portions of those allocations that are being used; it has taken back some of the unused portions). (Saunders and Wenig 2007, 128)

Thus with regard to human impacts, the flow and health of the SSR is determined largely by the policies of one provincial government (Alberta), although the SSR is an interprovincial river. Because the Alberta licences confer property rights, they represent a significant constraint on future adaptation to drought or water shortages (Hurlbert 2006b, 17).

Although irrigation is the largest consumptive use, municipal water supplies are another major use. (In contrast to Europe, most Canadian cities draw public water supplies from surface water.) The cities of Calgary, Red Deer, Lethbridge, Medicine Hat and Saskatoon are all on the SSR or one of its tributaries. (Three million people live in the Saskatchewan River basin.) Although some of these cities are rapidly growing, urban areas tend to be net contributors to the watershed, because wastewater is returned to the river, as is stormwater runoff (Partners 2009a, 13). In the event of seriously depleted flows in SSR, the city of Saskatoon, Saskatchewan, which is the farthest downriver, would feel the greatest impact.

Canadian Interprovincial Water Allocation Regime

Whereas the American regime for interstate water allocation is formal and highly legalistic, the Canadian regime for transboundary water allocation is far less formal, has almost no judicial involvement and is virtually invisible to the public. As of 2013 there were two agreements on interprovincial/territorial rivers. Of these, the Prairie Provinces Water Board (PPWB) (which deals with the SSR) is the oldest and most developed. An intergovernmental agreement on the Mackenzie River was signed in 1997, between the Northwest Territories, Alberta, Saskatchewan, British Columbia, the Yukon Territory, aboriginal groups and the Canadian federal government. The Mackenzie agreement itself is fairly hollow, largely a framework for bilateral substantive agreements to be negotiated later. It contains a toothless dispute resolution process. The negotiations lasted 25 years.

Actual water management and allocation are supposed to be set out in a series of seven bilateral agreements which '... will identify the share of water each jurisdiction has entitlement to, how this quantity will be defined over space and time and how it will be measured and reported' (Mackenzie River Basin Board 2009). Much to the aggravation of the downstream Northwest Territories, very little has happened since 1997 (Henton 2008, A1). The Northwest Territories is particularly concerned about several aspects of Alberta's water use (Partners 2009b, 34). As of 2013, the proposed joint management plan had yet to be created (Moore 2012).

Only one of these seven bilaterals has been negotiated (North West Territory and Yukon, in 2002) (Saunders and Wenig 2008, 130). The six remaining bilateral agreements are intended to manage four watersheds that feed into the Mackenzie, each of which was supposed to have agreements in place by 2014 (Mackenzie River Basin Board 2009). A single watershed may be governed by as many as five bilateral agreements. The Peace/Athabasca/Slave Watershed is to be governed by four bilateral agreements. Alberta and British Columbia have been negotiating over the Peace River since at least 1992 (Ellis 1992, A7). In 2005 these provinces signed a Memorandum of Understanding to negotiate the Bilateral Water Management Agreement; as of 2013 no such agreement had been reached (Government of British Columbia 2005).

The 1997 agreement created the Mackenzie River Basin Board (MRBB). In the event of a dispute between the parties, the parties can:

> refer an unresolved dispute, difference or question to the Mackenzie River Basin Board and request the Board to examine and report upon the facts and circumstances. The Board may appoint one or more experts to assist with resolving the dispute.

There is nothing to compel a signatory to comply with the MRBB's recommendation. Furthermore, any party to the Mackenzie River agreement may withdraw upon giving one year's notice. Thus under this condition, it is difficult to imagine the River Basin Board ever constraining the actions of the parties to the agreement.

The only functioning interprovincial water allocation agreement is the Prairie Provinces Water Board (PPWB), the focal point for allocating rivers flowing between Alberta, Saskatchewan and Manitoba. Although the PPWB is decades old, it remains so obscure it doesn't even rate a mention on Wikipedia. The agency seems content to remain invisible rather than transparent. Although it is a public agency, its annual reports have only been available online since 2009; accessing these documents online requires consenting to a set of terms and conditions (Prairie Provinces Water Board 2012).

Perhaps the PPWB has been obscure because it is/has been unnecessary. This is not the case. Not only is the allocation of water between the Prairie provinces likely to be more conflictual in the future, it was conflictual in the past, resulting in changes to the Prairie provinces' water governance.

Discussions on interprovincial water sharing began in the 1920s (Partners 2009b, 34). The PPWB originated from a 1930 interprovincial body, Western Water Board Agreement (WWBA), which simply kept records of water withdrawals (Stutt 1996, 73–4). This was when water supply was considered ample, prior to the advent of large-scale irrigation and when the cities on the Saskatchewan River were small.

The WWBA's successor, the Prairie Provinces Advisory Water Board focused more explicitly on water supply management. In 1947, after initial hesitation, the provincial governments permitted federal participation, in exchange for federal funding. Conflicts arose however, with the federal government favouring water allocation based on national priorities, while the provinces favoured their individual priorities. Initial proposals provided for a Water Board with sufficient autonomy to authoritatively resolve disputes arising between provinces.

The final form of the PPWB was agreed upon in 1948. It was designed as a planning body not a regulatory agency, tasked with conducting studies leading to a master water allocation plan. The initial mandate of the PPWB was limited to technical matters. As work proceeded, many technical issues turned out to be highly political. For example should flows used in the generation of hydroelectricity require formal allocation permits as consumptive uses (e.g. irrigation) do? In 1949 the PPWB met to develop general principles of allocation, assigning priorities to different projects, with preference given to already-existing projects over potential ones, consistent with FITFIR. As game theory would predict, discussions about how shortfalls in river flow should be allocated were particularly contentious among the parties.

The cornerstone of the PPWB is the 1969 Master Agreement on Apportionment. It provides that, at the Alberta/Saskatchewan border, Saskatchewan should receive 50 per cent of the natural flow of any transboundary river. (Similarly, at the Saskatchewan/Manitoba border, Manitoba should receive 50 per cent of the natural flow.) Flow is calculated on an annual, not monthly, basis. However the agreement also mentions 'equitable apportionment,' which implies considerations of timing of flows, not simply annual quantities (Bruce et al. 2003).

Most years Saskatchewan receives flows in excess of its entitlement under the agreement. However these provisions do not apply under drought conditions. Under the agreement, Alberta is entitled to a 2,100,000 million acre-foot minimum net annual depletion as long as the depletion does not reduce the flow at the boundary to less than 42.5 m^3/s. Alberta must also maintain a minimum daily discharge at the boundary of 42.5 m^3/s or one-half the natural flow, whichever is less. For example in 2001,

> … recorded flows in May, June and July were less than 50% of the monthly natural flow. However, as the flow at the boundary did not drop below 42.5 m^3/s, there was no violation of the Master Agreement. In effect, excess deliveries had been 'banked' earlier in the year; at the end of the year about 58% of the natural flow had been delivered. (Bruce et al. 2003, 98)

Arguably in the event of severe drought, Alberta would still be in compliance with the agreement even if flows fell below 42.5 m^3/s, provided half of the natural flow was being supplied.

There is no indication that these minimum flow requirements were developed on the basis of the ecological requirements of the river. For example, delivery requirements have been defined in such a way that the SSR and Red Deer River are treated as a single river. Thus, at Alberta's discretion, its delivery requirements can be met entirely by the flows of the SSR or the Red Deer River, or some combination of the two. While this may work for accounting purposes, it shows no consideration for the ecological health of each river basin (Figliuzzi 2002, 3).

The PPWB has experienced drought conditions many times since the agreement was negotiated. The PPWB only began apportionment monitoring in 1985. Between 1985 and 2000, deficits in delivery to Saskatchewan occurred in five years or one third of the time: 'Since the inception of interprovincial apportionment monitoring in 1985, deficits in delivery to Saskatchewan have occurred in 1988, 1989, 1992, 1998, and 2000' (Environment Canada 2003, 5). This conflicts with figures reported by Alberta Environment, indicating that no delivery deficits to Saskatchewan occurred between 1975 and 1997 (Figliuzzi 2002, 12).

ASSESSMENT OF CANADA'S REGIME

Integrated Water Resources Management

The Canadian regime for interjurisdictional water allocation shows no purposeful consideration of instream uses of water or groundwater/surface water interactions. The existing provincial system of water rights is not conducive to Integrated Water Resource Management (IWRM) and is not helpful in adapting to conditions of drought. A report by water experts notes:

> In [Western Canadian water] law, the most senior water right must be satisfied in any given year before other more junior rights are satisfied. This approach does not promote conservation and favours early uses over more beneficial uses. As well, since it licenses withdrawals from a stream, it does not set aside flows for ecological needs or needs such as waste assimilation. Although current laws provide a little more flexibility and there is a tendency for licensees to 'share the pain' during low flow conditions, problems with water shortages under the present water allocation system may lead to proposals to increase supply through importation from other basins, with potential environmental consequences to donor and recipient streams. (Bruce, et al. 2003, 98)

The primary emphasis is on allocation for consumptive uses, of which irrigation is the most significant. There is no basin-wide planning, nor stakeholder consultation, in interprovincial water governance. The province of Alberta has had extensive stakeholder consultation on basins *within* the province but without regard to the province's transboundary water obligations (Rosenberg International Forum on Water Policy 2007, 8–9). Thus the Canadian regime for transboundary rivers falls well short of the ideals of Integrated Water Resource Management.

Dispute Resolution

In theory the Canadian federal government has constitutional jurisdiction over interprovincial waters (Hurlbert 2006a). It also has jurisdiction over fisheries, which give it indirect powers to regulate water flows. In practice, however,

> ... the federal government has typically stressed deference to the interests of the province as resource owner, and it has preferred to work toward cooperative solutions (even if these are long in coming and not entirely satisfactory) over insistence on its constitutional authority. (Saunders and Wenig 2007, 127)

The Canadian regime does not appear well equipped to deal with the possibility of increased contention over interprovincial/territorial water allocation. As noted above, the framework agreement for the Mackenzie River has very weak dispute settlement provisions.

Although the longevity and consensus basis of the PPWB have been praised, the Board might not survive a crisis over water allocation, particularly for the South Saskatchewan basin. The PPWB provides for federal courts to be used to resolve disputes. This has yet to occur in the decades since the Master Agreement on Apportionment was negotiated. The possibility of unilateral withdrawal from the Master Agreement also limits the likelihood that the PPWB could ever compel a tough decision on allocation, or that such a decision could be enforced (Bruce et al. 2003, 159). Because there is no case law in this area, there is no indication how the federal government might intervene in an interprovincial water dispute.

Fairness

In Canada, the Harmon Doctrine of territorial sovereignty seems alive and well. The kinds of balancing tests for allocation found in US law and at least alluded to in EU principles and international law receive only passing mention. In the past, ample water supplies in most interprovincial watersheds made questions of rights in interprovincial allocation moot. In several interprovincial basins, downstream jurisdictions have no guarantees with regard to quantity or minimum flows of water.

Water law experts Owen Saunders and Michael Wenig (2007, 127) concluded that:

> in the absence of any serious indication on the part of the federal government that it will intervene in interjurisdictional disputes, upstream jurisdictions will either delay negotiations or only agree to the most modest undertakings, so as not to constrain their own future uses of interprovincial waters.

This prediction is consistent with the delay and deadlock observed in negotiations of bilateral agreements that are supposed to set out allocations within the Mackenzie watershed.

Even in the case of the Saskatchewan River, which is subject to an interprovincial agreement, the terms of the agreement guarantee a minimum quantity of water to Alberta, the upstream jurisdiction, not the downstream jurisdiction. The PPWB apportionment agreement only guarantees that Saskatchewan will receive at least 42.5 m^3/s, provided that represents less than half of the natural flow of the river. No figures are provided to determine the extent to which this flow is adequate for Saskatchewan's needs, or instream ecological values. There is no provision under the PPWB Master Agreement for compensation for deficits in delivery, in contrast to the 1909 Boundary Waters Treaty between Canada and the US (Figliuzzi 2002, 2).

Good Water Governance

The governance system for the Saskatchewan River falls far short of the ideals of MLG governance and basin governance. Under the vision of MLG, decisions are made at a multitude of points by several levels of government, incorporating the participation of many non-state actors. By contrast, the PPWB is executive federalism, where provincial sovereignty takes priority and the Canadian federal government takes a hands-off attitude. The PPWB is not particularly transparent and clearly sees government departments as its main clients, not citizens or stakeholders. While there has been a consultative stakeholder group for the watershed since 1993 (The Partners for the Saskatchewan River Basin), it is not clear that this group (or other watershed groups) has any connection to decisions about the basin (Rosenberg International Forum on Water Policy 2007, 8). Thus we are a long way away from the multilevel governance vision of policy making which is not state centric, not dependent on government authority.

The Mackenzie River Agreement's proposed structure of multiple bilateral agreements represents the antithesis of integrated watershed management, built on basin level governance. Governing separate watersheds by means of multiple bilateral agreements, each of which covers several watersheds, seems a sure fire guarantee of deadlock. It took 25 years to conclude the very vague Mackenzie River agreement, which left the controversial issues to be resolved by seven bilateral interprovincial/territorial agreements covering four watersheds. Given the glacial pace since negotiations began in the 1970s, it seems very optimistic to expect several agreements covering three contentious watersheds to be negotiated by 2014.

Canada Summary

The Canadian regime for interprovincial water allocation is more similar to an international agreement than, for example, a US interstate compact. Only one substantive agreement has ever been negotiated (the PPWB). It favours the upstream jurisdiction and, while largely complied with over the last 40 years, gives no explicit consideration to allocation for ecological benefits and in-stream uses. Although the federal government is, in theory, the arbiter of interprovincial water disputes, it has chosen not to get involved and appears content to leave transboundary water governance to interprovincial negotiation. The inflexibility of Alberta water licences suggests that under conditions of decreased river flows due to climate change there is a possibility of vigorous contention over interprovincial flows, particularly in the South Saskatchewan River. There is to date no body of Canadian law to guide adjudication of a water allocation dispute.

CONCLUSIONS

Although none of these regimes for transboundary river allocation is ideal, the American one is the only one that guarantees downstream states (and users) that their needs will receive equal consideration with those upstream. Given that downstream jurisdictions cannot force those upstream to provide water (but those upstream can readily over consume), it is the downstream jurisdictions' rights that are most in need of protection. For example, the downstream provinces and territories in the Mackenzie River basin are afforded no protection under the 1997 agreement and have no power to force the upstream provinces to even negotiate bilateral allocation agreements.

While the US system may encourage lengthy litigation, at least claimants know that decisions rendered are authoritative and enforceable. Rights for downstream parties will be upheld and compensation may be mandated. This reduces uncertainty for users. Dispute resolution provisions in the EU and in Canada are underdeveloped and/or untested. While international law should, in theory, resolve these disputes using the principle of equitable allocation, it is not clear what following this principle means in practice. Furthermore, there is no guarantee that a decision under international law will be implemented or enforced. In the Canadian case, recourse to federal courts should be possible but this has yet to occur so it is difficult to predict what the outcome would be.

None of the regimes reflects stakeholder-based governance. The EU mandates public consultation but this is not the kind of deliberative democracy considered preferable for Integrated Water Resource Management. In Canada and the US, the system is largely one of executive federalism (supplemented by court decisions) and similarly offers little real public involvement in decision making.

With regard to preserving ecological values, the EU and United States have provision for recognizing the ecological value of leaving water in the river, not allocating the entirety of a river's flow. Not all compacts recognize instream uses however, and it is too early to tell if the ecological aspects of the EU Water Framework Directive will be fully implemented. The Canadian system makes no explicit provision for protecting river flows for habitat or other non-consumptive uses, a serious shortcoming for a water management regime facing climate change and growing demand.

The American system has been criticized for the lengthy process of negotiation of and litigation over water compacts. The Canadian system's informality has been touted as an advantage. It does not however offer much advantage in timeliness. Negotiations over the Mackenzie are approaching the half-century mark, without achieving basin-wide governance. Furthermore, process considerations should not trump differences in substantive outcomes between the two systems. The EU's regime, based on the Water Framework Directive, may prove superior in the end. Because it is based in legislation, it can be authoritative and enforceable but provides the possibility of ongoing revision without being held hostage by any individual jurisdiction.

NOTES

1. It is not a Prisoner's Dilemma game, which is not a zero sum game because a win-win solution exists.
2. Jerome Muys (2003) provides the higher figure of 26 compacts but does not list them.

REFERENCES

Albiac, Jose, Michael Hanemann, Javier Calatrava, Javier Uche, and Javier Tapia (2006), 'The rise and fall of the Ebro water transfer,' *Natural Resources Journal* 46, 732–33.

Ansink, Erik and Arjan Ruijs (2008), 'Climate change and the stability of water allocation agreements,' *Environmental Resource Economics* 41, 249–66.

Bruce, J.P., H. Martin and P. Colucci (2003), 'Climate change impacts on boundary and transboundary water management,' *A Climate Change Action Fund Project. Natural Resources Canada*, Project A458/402, available at http://www.saskriverbasin.ca/file/Final%2520Report%2520A458-402%2520CCAF.pdf (accessed 22 May 2013).

Das Gupta, A. (2008), 'Implication of environmental flows in river basin management,' *Physics and Chemistry of the Earth* 33, 298–303.

del Castillo-Laborde, Lilian (2009), 'Case law on international watercourses,' in J.W. Dellapenna and J. Gupta (eds), *The Evolution of the Law and Politics of Water*, Springer Science + Business Media B.V. pp. 319–35.

Draper, Stephen E. and James E. Kundell (2007), 'Impact of climate change on transboundary water sharing,' *Journal of Water Resources Planning and Management* 133, 405–15.
Ellis, Erin (1992), 'Deal with B.C. on river flows could be signed this summer,' *Edmonton Journal*, 18 March, p. A7.
Environment Canada (2007), 'Water demand and availability indicator based on a 30 years long term yearly average water supply and 2005-2007 water use,' available at http://www.ec.gc.ca/eau-water/default.asp?lang=En&xml=C982BC4F-A2B1-49B8-96E6-006E98B4FAA8 (accessed 28 May 2013).
Environment Canada (2003), *Annual Report Canada Water Act 2000-2001*, available at http://www.ec.gc.ca/Publications/89D6EB74-CB38-4CA5-895F-CA4A0AB99500%5CThe-Canada-Water-Act-Annual-Report_-2000-2001.pdf (accessed 20 May 2013).
European Commission (2013), 'River Basin Management Plans 2009-2015 – information on availability by country,' available at http://ec.europa.eu/environment/water/participation/map_mc/map.htm (accessed 23 May 2013).
European Commission (2012a), 'The role of water pricing and water allocation in agriculture in delivering sustainable water use in Europe: Final report,' available at http://ec.europa.eu/environment/water/quantity/pdf/agriculture_report.pdf (accessed May 24 2013).
European Commission (2012b), 'Report from the Commission to the European Parliament and the Council on the implementation of the Water Framework Directive (2000/60/EC) River Basin Management Plans' COM (2012) 670 final, available at http://ec.europa.eu/environment/water/water-framework/pdf/COM-2012-670_EN.pdf (accessed 27 May 2013).
European Commission (2012c), 'European Overview Accompanying the Document Report from the Commission to the European Parliament and the Council on the Implementation of the Water Framework Directive (2000/60/EC) River Basin Management Plans,' SWD (2012) 379 final.
European Commission (2008), 'Report from the Commission to the Council and the European Parliament: Follow up Report to the Communication on Water Scarcity and Droughts in the European Union,' COM (2007) 414 Final,' COM (2008) 875 final.
European Commission (2007), 'Addressing the Challenge of Water Scarcity and Droughts in the European Union,' COM (2007) 414 final.
European Commission, DG Environment (2012), 'EU River Basin Districts indicating transboundary co-operation,' available at http://ec.europa.eu/environment/water/water-framework/facts_figures/pdf/Transboundary-cooperation-%202012.pdf (accessed 27 May 2013).
European Commission, DG Environment (2008), 'Water note 1. Joining forces for Europe's shared waters: Coordination in international river basin districts', available at http://ec.europa.eu/environment/water/participation/pdf/waternotes/water_note1_joining_forces.pdf (accessed 22 May 2013).
European Council (2009), 'Council Regulation (EC) No 73/2009 of 19 January 2009 Establishing Common Rules for Direct Support Schemes for Farmers under the Common Agricultural Policy.'

European Environment Agency (2009a), 'Water Exploitation Index,' available at http://www.eea.europa.eu/data-and-maps/figures/water-exploitation-index-wei-2 (accessed 23 May 2013).
European Environment Agency (2009b), *Water Resources across Europe – Confronting Water Scarcity and Drought*, EEA Report No 2/2009, Copenhagen: European Environment Agency.
Figliuzzi, Sal (2002), 'South Saskatchewan River sub-basin contributions to international and interprovincial water-sharing agreements,' Edmonton: Alberta Environment, available at http://ssrb.environment.alberta.ca/pubs/SSRB_Contributions_Agreements.pdf (accessed 29 May 2013).
Garrido, Alberto and M. Ramón Llamas (2009), 'Water management in Spain: An example of changing paradigms,' in Ariel Dinar and Jose Albiac (eds) *Policy and Strategic Behaviour in Water Resource Management,* London: Earthscan, pp. 125–44.
Gleick, Peter (ed.) (2000), *Water Conflict Chronology*, available at http://www.worldwater.org/conflictchronologychart.PDF (accessed 21 May 2013).
Gleick, Peter (2009a), 'Water wars – Myth or reality,' *SFGate*, available at http://www.sfgate.com/cgi-bin/blogs/gleick/detail?blogid=104&entry_id=40585 (accessed 21 May 2013).
Gleick, Peter (2009b), 'Whose water is it? Water rights in the age of scarcity,' 2 August. Available at http://www.sfgate.com/cgi-bin/blogs/gleick/category?blogid=104&cat=2624. (accessed 13 August 2009).
Goodman, Brenda (2007), 'Drought-stricken South facing tough choices,' *New York Times,* 16 October.
Government of British Columbia (2005), *British Columbia – Alberta Memorandum of Understanding – Bilateral Water Management Agreement Negotiations*, available at http://www.gov.bc.ca/igrs/attachments/en/05mar18_water_mgmt_neg_mou.pdf (accessed 10 August 2009).
Grant, Douglas L. (2003), 'Interstate water allocation compacts: When the virtue of permanence becomes the vice of inflexibility,' *University of Colorado Law Review* 74, 105–81.
Hanada, A. P. and L. Hancock (2004), 'The Pecos River case,' in T. Pearson d'Estrée and B.G. Colby (eds), *Braving the Currents: Evaluating Environment Conflict Resolution in the River Basins of the American West.* Boston: Kluwer Academic Publishers, pp. 157–76.
Henton, Darcy (2008), 'Territory demands action on water; Existing deal languishing: Minister,' *Edmonton Journal.* 17 August, p. A1.
Holwarth, Fritz (2002), 'The EU Water Framework Directive – a key to catchment-based governance,' *Water Science and Technology* 45(8), 105–12.
Hurlbert, Margot (2006a), 'Interjurisdictional Water Law – SSRB,' IACC Project Working Paper No. 26, available at http://www.parc.ca/mcri/pdfs/papers/iacc026.pdf (accessed 22 May 2013).
Hurlbert, Margot (2006b), 'Water law in the South Saskatchewan River basin (Alberta/Saskatchewan),' IACC Project Working Paper No. 27, available at http://www.parc.ca/mcri/pdfs/papers/iacc027.pdf (accessed 22 May 2013).
Louka, Elli (2008), *Water Law and Policy: Governance without Frontiers.* New York: Oxford University Press.

Mackenzie River Basin Board (2009), *Bilateral Water Management Agreements: Guidance Document*, available at http://www.mrbb.ca/uploads/files/general/28//mrbb-bilateral-guidance_document-final-1.pdf (accessed 21 May 2013).
Moore, Dene (2012), 'Rescuing the Mackenzie and all it feeds,' *Canadian Press*, 5 September, available at http://thetyee.ca/News/2012/09/05/Mackenzie-River-Basin/ (accessed 5 September 2012.)
Mostert, Erik (2001), *Case Study: International Co-operation in the Scheldt and Meuse River Basins*, available at http://www.gwptoolbox.org/images/stories/cases/en/cs%2028%20transboundary.pdf (accessed 21 May 2013).
Mostert, Erik (2003), *Conflict and Cooperation in the Management of International Freshwater Resources: A Global Review.* Paris: UNESCO.
Muys, Jerome, (2003), 'Interstate water management in the Colorado River Basin in the United States', paper presented at the World Bank Conference on 'Water Development', available at http://siteresources.worldbank.org/EXTWSS/Resources/337301-1147283814231/2532554-1153756214980/Jerry Muys-Paper.pdf (accessed 10 August 2009).
Mylopoulos, Yannis A., E. Eleftheriadou, E.Kampragou (2004), 'The transboundary catchment of River Nestos and the bilateral agreement between Greece and Bulgaria,' The Open TRANSCAT Conference on Integrated Water Management of Transboundary Catchments available at http://www.feem-web.it/transcat_conf/conf_papers/Eleftheriadou.pdf (accessed 23 May 2013).
Oregon State University (2007), *International Freshwater Treaties Database*, available at http://www.transboundarywaters.orst.edu/database/interfreshtreat data.html (accessed 10 August 2009).
Prairie Provinces Water Board (2012), *Annual Reports*, available at http://www.ppwb.ca/documents/47/index.html (accessed 13 May 2013).
Rosenberg International Forum on Water Policy (2007), Report of the Rosenberg International Forum on Water Policy to the Ministry of Environment, Province of Alberta, available at http://www.waterforlife.gov.ab.ca/docs/Rosenberg_Report.pdf (accessed 13 May 2013).
Ruhl, J.B. (2003), 'Equitable apportionment of ecosystem services: New water law for a new water age,' *Journal of Land Use* 19(1), 47–57.
Ruhl, J. B. (2005), 'Water wars, Eastern style: Divvying up the Apalachicola-Chattahoochee-Flint River basin,' *Journal of Contemporary Water Research and Education* 131, 47–54.
Saunders, Owen and Michael W. Wenig (2007), 'Whose water? Canadian water management and the challenges of jurisdictional fragmentation,' in Karen Bakker (ed.) *Eau Canada: The Future of Canada's Water*, Vancouver: UBC Press, pp. 119–42.
Schindler, W. (2001), 'The cumulative effects of climate warming and other human stresses on Canadian freshwaters in the new millennium,' *Canadian Journal of Fisheries and Aquatic Sciences* 58, 18–29.
Schlager, Edella and Tanya Heikkila (2009), 'Resolving water conflicts: A comparative analysis of interstate river compacts,' *Policy Studies Journal* 37(3), 367–92.
Sherk, George W. (2005), 'The management of interstate water conflicts in the twenty-first century: Is it time to call uncle?' *New York University Environmental Law Journal* 12, 764–828.

Snowden, Benjamin L. (2005), 'Bargaining in the shadow of uncertainty: Understanding the failure of the ACF and ACT compacts,' *New York University Environmental Law Journal* 13, 134–97.

Stutt, Roderick Irwin (1996), 'Water policy-making in the Canadian plains: Historical factors that influenced the work of the Prairie Provinces Water Board (1948-1969)', doctoral dissertation, University of Regina, Saskatchewan, Canada.

The Partners for the Saskatchewan River Basin (2009a), *From the Mountains to the Sea: The State of the Saskatchewan River Basin*, available at http://www.saskriverbasin.ca/file/SRB%20Summary.pdf (accessed 12 August 2009).

The Partners for the Saskatchewan River Basin (2009b), *Institutional Arrangements*, available at http://www.saskriverbasin.ca/file/SRB%20CH03%20Institutional.pdf (accessed 12 August 2009).

'Top court won't hear tri-state water dispute,' (2012) *Gwinnett Daily Post*, 25 June, available at http://www.gwinnettdailypost.com/news/2012/jun/25/top-court-wont-hear-tri-state-water-dispute (accessed 18 September 2012).

United Nations Environment Program (2009), *Hydropolitical Vulnerability and Resilience Along International Waters: Europe*, Nairobi: UNEP, available at http://www.unep.org/dewa/Portals/67/pdf/europe.pdf (accessed 20 May 2013).

US Interstate Freshwater Compacts Database, available at http://ocid.nacse.org/tfdd/domesticCompacts.php (accessed 15 July 2010).

Zimmerman, Joseph F. (2012), *Interstate Water Compacts: Intergovernmental Efforts to Manage America's Water Resources*. Albany, NY: State University of New York Press.

PART III

Multilevel governance of climate change mitigation

6. Climate governance in the European Union multilevel system: the role of cities

Kristine Kern

Today it is widely acknowledged that cities have become important players in governing climate change in the European Union (EU). Since climate change is a complex problem that requires action at local and regional levels, the success of EU climate change policy depends ultimately on subnational authorities. As an increasing number of citizens live in cities, a high proportion of greenhouse gas (GHG) emissions are generated at the local level. Thus many European cities started climate mitigation initiatives to reduce their GHG emissions as early as the 1990s. Despite these efforts, the effects of climate change have become noticeable at the local level and have triggered the development of local climate adaptation strategies. However cities are not only sites where climate change causes severe risks to natural and social systems; they are also the places where the social and technological innovations that help to cope with the new challenges are generated. Many European cities have set GHG emission targets, measure and report their emissions annually, have completed risk assessments, developed adaptation plans and have extended their reach with voluntary agreements (Carbon Disclosure Project, 2012).

As European cities are strongly affected by EU climate change policy, this chapter concentrates on climate governance in the EU multilevel system. It is assumed that the implementation of EU climate change policy at local level changes local policies and, vice versa, local climate governance influences the development of EU climate policy. This assumption is supported by the fact that cities have opened offices in Brussels, founded transnational city networks such as the Climate Alliance and signed the 'Covenant of Mayors,' an initiative of the EU Commission which aims to reduce the GHG emissions of cities by more than 20 per cent and establish direct relations between the EU Commission and European cities.

The main goal of this chapter is to analyse the role cities play in EU multilevel climate governance. In the first place, multilevel climate governance implies that EU programs are eventually implemented at local and regional levels. This top-down approach has shaped the debate on climate governance in the EU ('hierarchical climate governance'). Second, cities have tried to alter the effects of hierarchical climate governance by establishing direct links with EU institutions. This perspective combines top-down with bottom-up approaches ('vertical climate governance'). Third these types of climate governance need to be complemented by a third type because cities have developed various tools to facilitate knowledge transfer among cities ('horizontal climate governance'). Although EU institutions play either no role here, or merely a facilitative one (in particular through project funding), it is essential for EU climate governance. Thus EU climate governance is a combination of these three types and the specific combination determines the outcome of EU climate policy.

The next section provides a general overview of the debate on multilevel climate governance and examines how cities can be integrated into this approach. Subsequently, the chapter discusses the different forms of climate governance: hierarchical climate governance; vertical climate governance; and horizontal climate governance. The final section summarizes the discussion and draws some conclusions about the role cities play in EU climate governance.

CITIES IN EU CLIMATE GOVERNANCE

In terms of the discussion of the role of cities in EU climate governance, their position in the multilevel structure of the European Union would appear to be crucial. The multilevel governance approach starts from the assumption that the EU has become a system with multiple levels or spheres of governance, including European, national and subnational arenas (Hooghe and Marks 2003; Bache and Flinders 2004; Smith 2007; Bache 2008; Monni and Raes 2008; Gustavsson et al. 2009; OECD 2010; Westerhoff et al. 2011). Multilevel governance includes the whole range of institutional arrangements that supply order, including hybrid and private forms of governance.

Originally, the multilevel governance approach was developed for analysing EU regional policy (Hooghe and Marks 2003). However, it is not limited to studying the EU multilevel system, but can be applied to all political systems with multiple levels or spheres of government (Enderlein et al. 2010). The multilevel governance approach emphasizes that the boundaries between different levels of government have become blurred. Authority is not simply shifted upwards and downwards (Rosenau 2003); it has also become dispersed among a variety of public and private actors.

Multilevel governance approaches differ from classical approaches to European integration in at least two respects: first, this approach does not consider the levels of government as parts of a stable hierarchical arrangement. Instead multilevel governance approaches assume that competencies between local, national and European institutions are shifted not only upwards to the European Union but also downwards from the nation states to regions and cities (Rosamond 2007; Rosenau 2003; Pierre and Peters 2000, 77). In a multilevel system it becomes difficult to determine the boundaries between different policy arenas because actors pursue multilevel strategies such as venue shopping and may start parallel initiatives at different levels. Second, this approach is not limited to state-based forms of regulation, but acknowledges the importance of subnational and nonstate actors. Hybrid arrangements of diverse actors that combine different approaches to governance seem to be more effective than alternative forms of governance (Delmas and Young 2009, 8).

Figure 6.1 shows the three forms of climate governance in multilevel systems. In hierarchical climate governance, authority is concentrated at the federal level, which has exclusive competences for decision making, while lower levels of government execute policies. This type of climate governance is characterized by top-down relations among the different levels of government. 'Those governing are, or see themselves, as in some way superimposed above those governed' (Kooiman 2003, 115). From this perspective, cities execute decisions made at EU and national level. Hierarchical climate governance requires strong governments at EU and national level with the authority, power and capacity to harmonize policies by steering and controlling subnational authorities (cf. Kooiman and Jentoft 2009). In hierarchical climate governance, governmental actors play a dominant role, while the influence of non-government organizations (NGOs), the business sector and local authorities on policy making and implementation is limited. However, this mode of governance has consequences for the effectiveness of policies because hierarchical governance requires a comprehensive system of controls and sanctions. The legitimacy of this mode of climate governance rests on decision making at EU and Member State levels, while cities have little discretion when implementing these decisions.

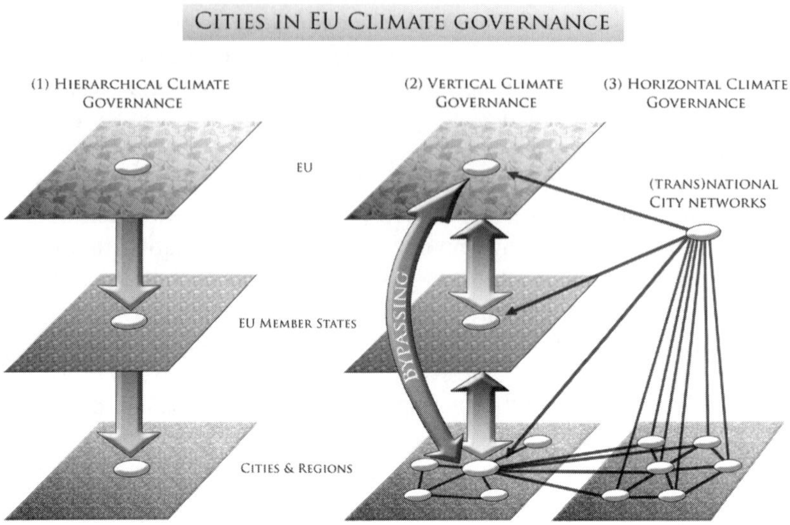

Figure 6.1 Cities in EU climate governance

Source: author

For the EU, hierarchical climate governance plays a limited role because in most Member States local climate change policy is a voluntary task and direct relations between the EU and cities do not exist. By contrast, vertical climate governance is characterized by a flexible institutional design and interdependent relations between levels. Lower levels of government have enough authority to influence decision making at higher levels through formal and informal networks. In other words top-down relations between governmental levels are complemented by bottom-up relations. This is in line with Elinor Ostrom's (2010, 552) argument that polycentric systems, i.e. systems with multiple governing authorities at different scales, have considerable advantages due to their mechanisms for mutual monitoring, learning and adaptation.

The vertical interplay between institutions creates new options for cities to interact directly with EU institutions and bypass the Member States (see Figure 6.1). Hierarchical governance may still play a role here, but it is superseded by interdependent relations among the authorities at various levels of the governance system. This implies that competences are shared between governments at different levels and top-down decisions are replaced by joint decision making within formal and informal networks. Although relatively centralized forms of governance can be found in certain areas of EU environmental policy, there are also relatively decentralized policy areas.

Pioneering Member States try to upload policy innovations to the European level since this strategy helps to avoid a misfit between national and EU legislation (cf. Börzel and Risse 2007). When such strategies are successful, other Member States that are less innovative and active at the EU level are confronted with a misfit between national and EU legislation. This may affect cities because EU legislation needs to be transposed in national law and implemented by the Member States.

The effectiveness and legitimacy of vertical climate governance depends on formal and informal forms of inclusion of subnational authorities and nonstate actors in decision making and implementation. This is most obvious within nation states when national and subnational governments cooperate in formalized ways. It can be assumed that vertical climate governance will be more effective and legitimate than hierarchical climate governance if the positions of subnational authorities are taken into account before final decisions are taken.

Horizontal climate governance has been neglected by the multilevel governance debate. On the one hand, this is not surprising because this third type of climate governance concentrates primarily on the horizontal relations between and among cities. On the other hand, it can be argued that the horizontal dimension of climate governance needs to be integrated in a comprehensive model of multilevel climate governance because horizontal climate governance has important repercussions for hierarchical and vertical climate governance.

For horizontal climate governance, the debates on policy diffusion, policy transfer, policy learning and policy mobility are most relevant (Kern 2000; Evans 2004; Rose 2005; Citi and Rhodes 2006; Bulmer et al. 2007; Karch 2007; Marsh and Sharman 2009; McCann and Ward 2012). Horizontal climate governance improves the prospects for knowledge transfer and learning among cities. In the European Union, three forms of networking can be distinguished: (i) bilateral networking between two cities; (ii) project networking of a limited number of cities that may be supported and funded through national or EU programs; and (iii) the multilateral networking of cities (Kern and Bulkeley 2009). All three types of networks can be restricted to cities in the same country or consist of cities that are located in different countries.

City networks can be regarded as a combination of vertical and horizontal climate governance because they have two main functions: (1) lobbying at national and at EU level; and (2) facilitating knowledge transfer and learning among their member cities (see Figure 6.1).

Transnational networking may strengthen the position of cities within the EU multilevel system because it creates transnational spaces that help to bypass and undermine hierarchical relationships. However the effectiveness and legitimacy of such arrangements seems to be rather limited because knowledge transfer requires voluntary initiatives. In addition networks lack the authority to sanction member cities which do not comply with the rules and norms set jointly by the members of the network (such as GHG emission reduction targets).

In multilevel systems it can be expected that hierarchical climate governance is substituted and complemented by vertical and horizontal climate governance. The specific combination of the three types of climate governance may change over time. Since this applies not only to the EU but also to the Member States, the combination of hierarchical, vertical and horizontal climate governance may also differ among the 27 EU Member States.

HIERARCHICAL CLIMATE GOVERNANCE IN THE EUROPEAN UNION

Cities are strongly affected by the implementation of EU policies. Most important in this respect are EU directives, for example in the area of internal market policies, which must be transposed into national law before they can be implemented at national and subnational level. The role which local authorities play in the European Union has changed considerably over time. Prior to the Single European Act (1986), European legislation had only limited effects on subnational governments. By the early 1990s, this situation had changed fundamentally and today there is almost no area of local or regional policy that is not affected by European regulation (Bourne 2007; Allen 2005; John 2000, 879). Thus cities are affected by many decisions which are made in Brussels. This includes market making policies, such as the liberalization of the energy markets and the introduction of the EU Emission Trading System, and market correcting policies such as the EU's regional policy, including the development of renewable energy capacities at regional level (cf. Sbragia 2004). Although most legislation that is implemented locally and regionally is the product of EU decision making, this influence is only indirect because it comes in the guise of national legislation.

Like other policy areas, climate change policy has become widely affected by EU multilevel governance. European climate change policy has been driven primarily by international developments, in particular the Rio Summit, which stimulated European, national and subnational initiatives. In Europe, a variety of policy initiatives at all government levels were developed at a relatively early stage. The EU has enacted a variety of policies with the aim of reaching its Kyoto target (such as the Renewable Electricity Directive, the Biofuels Directive and the Co-generation Directive). In March 2007, the European Union reached a compromise on its goals for the post-2012 period and decided to reduce GHG emissions by 20 per cent from 1990 levels by the year 2020. Other goals for 2020 include a binding target for meeting 20 per cent of total primary energy requirements with renewable energy sources and a 20 per cent increase in energy efficiency.

The EU already reached its common target (EU burden sharing agreement) of an 8 per cent GHG emissions reduction (base year 1990) for the first commitment period, but compliance in the Member States differs considerably. The EU encompasses countries which have already achieved a remarkable decline in GHG emissions (Germany, United Kingdom) and countries with increasing GHG emissions. However, all EU Member States are under considerable pressure to develop successful programs, including measures which directly target subnational governments. Although the EU burden sharing and effort sharing agreements implied specific GHG reduction targets for individual EU Member States, this did not lead to the introduction of hierarchical forms of climate governance that oblige regions and cities to start initiatives.

Although hierarchical relations between the EU and European cities do not exist, a few examples of hierarchical climate governance can be found at Member State level. In the United Kingdom (UK) for example, intergovernmental relations were relatively centralized, at least until recently. Cities in the UK could act only within the boundaries set by the national government, i.e. local decision making had to fulfil national demands (Bulkeley and Kern 2006; John and Copus 2011, 30). At the national level, the UK has been among the leading Member States in implementing the EU climate change programs, including relatively ambitious GHG reduction goals which have superseded the EU goals for the UK. These goals resulted in the development of local performance standards. In England, local authorities have been assessed on their climate change policy since 2008, based on three obligatory performance indicators: (1) CO_2 reductions from local authority operations; (2) per capita reductions in CO_2 emissions in the local authority area; and (3) adapting to climate change. This is part of a streamlined set of 198 national performance indicators for local authorities. These indicators were included in the Local Area Agreements (performance agreements for local authorities that were negotiated with central government)

(Fudge and Peters 2009). Comparable mandatory instruments for implementing EU and national climate policy have never existed in most other Member States and even in the UK the situation is now changing considerably – due to the Cameron government's policy of 'new localism' which will lead to far more discretion at the local level. Since in most Member States local authorities are relatively independent from the national government (for example in the Nordic countries and in Germany), Member States developed different policy instruments, such as subsidies, to influence local climate governance. This means that climate governance at both EU and at Member State level is dominated by vertical climate governance.

VERTICAL CLIMATE GOVERNANCE IN THE EUROPEAN UNION

European Union and national climate change policies can be implemented only if they are coordinated and integrated with regional and local approaches. Many leading European cities started their climate initiatives earlier and set more ambitious GHG reduction targets than their national governments and the EU. Thus local climate change initiatives have become an important part of national climate governance, in particular in those Member States with the most ambitious national programs. Although a considerable percentage of the population live in cities with relatively ambitious reduction goals, initiatives in the Member States are still needed to support local initiatives.

Vertical climate governance can be found within Member States but also at EU level. The relations within Member States as well as between EU institutions (in particular the EU Commission) and subnational authorities can take different forms. Most important are forms of direct lobbying to influence decision making and the development of collaborative programs between institutions at different levels. Such programs seem to be most successful when they are combined with funding programs for local authorities set up by national governments and the EU.

Examples in the Member States include experiences in the Netherlands and in Germany. In both countries, state-local relations are cooperative and the national government is in a relatively weak position in terms of its influence on the implementation of climate change policy at local level. In the past, Dutch climate policy at the local level has been based primarily on a covenant. The Dutch 'Klimaatcovenant' was a multilevel arrangement involving local government, provinces and several national ministries. The so-called BANS agreement (Bestuursakkord Nieuwe Stijl) started in 2002. In order to receive funding, local authorities had to undergo an initial scan and had to implement a comprehensive climate action plan using a common methodology. This scheme was based on a performance card ('Prestatiekaart'), which comprised targets, policies and measures, grouped according to the relevant fields of work (municipal facilities, housing, mobility, business, etc.) Cities could choose between three activity levels: active; advanced; and innovative. This program was very successful and resulted in several hundred local climate scans and more than 250 municipal implementation plans. In 2008 the BANS program was transformed into the SLOK program (Stimulering LOkale Klimaatinitiatieven), which was less effective than BANS because SLOK was rather a general subsidy program with fewer rules and obligations. The recently developed 'Klimaatagenda 2011-2014,' which is part of the National Roadmap Climate 2050, does not include a subsidy program at all (NOVEM 2003; VROM 2007; Gupta 2007; Gupta et al. 2007; Bommel and Kuindersma 2008; den Exter 2012). The Dutch case shows that the financial capacity of cities to set up climate protection programs depends not only on local factors but also on national preconditions. As the national government will not provide any financial incentives in the future it can be expected that most cities, in particular small and medium sized cities, will have problems to maintain their activities and reach their GHG reduction goals.

In Germany, an ambitious GHG emissions reduction strategy was introduced as early as 1990. Although it was ultimately not met, the national GHG reduction target of 25 per cent by 2005 was the highest target adopted among the pioneering countries in the EU. Until recently, the federal government's role for subnational climate policy was limited to the collection of information and the dissemination of knowledge on best practice cases. By 1997 the German federal government already provided the municipalities with guidelines for local climate protection ('Leitfaden Kommunaler Klimaschutz') (Deutsches Institut für Urbanistik, 2011). These guidelines contain sections on general aspects of local climate protection, steps for the development of local concepts, the exchange of experience, funding for climate protection activities and recommendations for all relevant sectors (energy, transport, urban planning and land use and waste). Such facilitating approaches appear to dominate in federal states such as Germany. In such political systems the states, for example the German federal state of North-Rhine Westphalia, constitute an additional layer of government, which may run more comprehensive programs for local authorities than the federal government. In federalist countries, national (funding) programs are often developed at a later stage (Kern et al. 2007). This may explain why the German federal government has only recently set up a subsidy program and a service agency for local climate protection as part of its national climate protection initiative (Servicestelle Kommunaler Klimaschutz). However, the German case also demonstrates that, even in the absence of regulation and subsidy, it is possible to create incentives for the institutionalization of climate change policy at the local level. Competitions and awards can be used to recognize the best performing local authorities. These forms of voluntary benchmarking help to generate more action on the ground.

Vertical climate governance differs from hierarchical climate governance in assuming that direct links between local authorities and EU institutions can be established and that lobbying is not necessarily restricted to the national realm. Many cities and regions in the EU have changed their attitudes and behaviour when confronted with European integration. European initiatives are now regarded more as new opportunities than as restrictions. The new opportunity structure draws cities to Brussels: they set up offices (Goldsmith 2003, 124; Kassim 2005, 307), cooperate with each other to strengthen their influence on EU decision making and apply jointly for funds.

Similar to the position of a national government in a federal state, in the past EU institutions did not have any direct links to European cities until the European Commission started initiatives to establish such links. While associations and transnational networks have increased their initiatives in Brussels, new EU initiatives, in particular initiatives launched by the EU Commission, have altered the opportunity structure for local and regional authorities and their representatives. The Commission published its 'White Paper on European Governance' in 2001 and the basis for direct cooperation between the EU and subnational authorities can be located in this document: 'At EU level the Commission should ensure that regional and local knowledge and conditions are taken into account when developing policy proposals' (European Commission 2001, 23–4). The Commission strategy can be explained by the fact that the EU has become more dependent on information and expert advice from civil society, business groups and subnational authorities (Ausschuss der Regionen, 2010).

It is now widely acknowledged that the EU has not only imposed new constraints on subnational authorities but that Brussels also offers them new opportunities and new channels of influence (Rechlin 2004; Kassim 2005, 303). European Union legislation made in Brussels is not simply handed down to the Member States' administrations and then implemented by local authorities. When cities begin to bypass national governments and start lobbying at EU level, while national governments limit their gatekeeper position, cities develop from policy takers to policy makers (Schulze 2003).

In contrast to hierarchical climate governance, which implies the transposition of EU legislation into national law and its implementation at the local level, vertical climate governance requires the institutionalization of foreign relations, first, in the city government (for example setting up an office for EU affairs). Second, the most active cities have established such specialized offices not only at home but also in Brussels (most often in cooperation with the neighbouring regions) and represent their interests independently of their national governments. Third, smaller cities with insufficient capacities to set up their own office in Brussels can also gain access to EU institutions through city networks that represent their interests in Brussels. These new institutional arrangements enable cities to pursue multilevel strategies and become active at regional, national, EU and international levels. Cities may even develop strategies to use their growing influence in Brussels to change the intergovernmental relations in their home countries.

Vertical climate governance is directly related to EU funding, such as the Intelligent Energy Europe (IEE) program, which was set up to facilitate the implementation of EU policies at sub-national level (for example by establishing local and regional energy agencies). It supports the development of clean and sustainable solutions and facilitates knowledge transfer among European cities. The three main objectives of the program are: (1) promoting energy efficiency and the rational use of energy sources; (2) increasing the use of renewable energy sources and encouraging energy diversification; and (3) stimulating energy efficiency and the use of renewable energy sources in the area of transport. The successful implementation of this funding scheme requires a multi-actor approach that involves a wide variety of actors, ranging from private companies to local and regional governments and from regional energy agencies to research institutions.

In January 2008 the European Commission, supported by the Committee of the Regions and the EU Parliament, initiated the 'Covenant of Mayors.' The initiative aims to bring together the mayors of Europe's pioneering cities to improve energy efficiency and promote cleaner energy production. This includes a formal commitment by the cities to reduce their CO_2 emissions by more than 20 per cent by 2020. The scheme is based on a voluntary agreement, but goes beyond the facilitation of local initiatives because it is linked to EU funding schemes (in particular the Intelligent Energy Europe program). This initiative is backed up by more than 100 national and sub-national 'covenant coordinators' (such as Italian provinces) and more than 70 'covenant supporters' (mainly national and sub-national associations and networks of cities and towns). Local authorities need to prepare a baseline emissions directory, present a sustainable energy action plan and provide regular implementation reports. By September 2012 almost 4300 cities with almost 168 million inhabitants had joined the initiative, around 1800 Sustainable Energy Action Plans had been submitted and 400 of these plans had been accepted. The European Commission set up a 'benchmark for excellence' mechanism. Benchmarks of excellence are local initiatives and programs which have been realized by the signatories and seem to be transferable from these cities to other cities. Cities which do not comply with the rules can be excluded from the plan. Moreover a Covenant secretariat, run by various city networks and funded through the Intelligent Energy Europe program, facilitates monitoring, networking and promotion.

Since the early 1990s we also find a rapid diffusion of the membership of European cities in three transnational city networks in the area of climate change policy (Climate Alliance, Cities for Climate Protection Campaign and Energy Cities) (see Figure 6.1).[1] These networks began as transnational organizations, based exclusively on transnational membership. Thus they differ considerably from traditional EU associations (such as the Council of European Municipalities and Regions) that are rather umbrella organizations of national associations with national membership (Niederhafner 2008). These institutionalized forms of networking enabled many European cities to actively pursue multilevel strategies and operate efficiently within the European multilevel governance context. Transnational city networks are backed by the EU because direct relations between the EU and sub-national actors can help the Commission to implement EU policies (Goldsmith, 2003, 121; Heinelt and Niederhafner 2008; Niederhafner 2008). Thus these networks established their own offices in Brussels and joined forces. Energy Cities for example, has not only run its own office in Brussels for many years, but since 2007 this network has also shared its office with the Climate Alliance.

The three transnational city networks that are active in the area of energy and climate policy in Europe have many shared characteristics. Their general goals are almost identical: they aim to (1) set voluntary GHG reduction goals for their member cities; (2) improve local capacities for climate mitigation and adaptation; (3) represent the interests of their member cities at national, European and international level; and

(4) support the exchange of experience and transfer of knowledge among their member cities (Kern 2001; Kern and Bulkeley 2009). These goals show that transnational city networks combine features of vertical climate governance and horizontal climate governance (see Figure 6.1).

However the mobilization of cities differs, first, among cities within the same country due to specific characteristics such as city size, capacities (especially financial capacities) and willingness to learn from cities abroad. In many countries, including Germany, the Netherlands and Sweden, cities have the option to join various global, European and national networks. While the most innovative cities may join several national and international networks, smaller cities may prefer to become active in national or regional city networks only. Second, the Covenant of Mayors shows that the mobilization of cities may also differ considerably along national lines because intergovernmental relations and national programs vary across the different Member States in the EU. Cities may get interested in EU initiatives such as the Covenant of Mayors because they expect better chances of getting financial support for their climate initiatives from the EU due to a lack of national initiatives. This may explain that the Covenant of Mayors has become most attractive for Italian and Spanish cities.

It appears that local authorities that have become active in the area of energy and climate governance have opened offices in Brussels, become members of transnational city networks, and participated in EU projects. They have joined the Covenant of Majors and are strongly committed to reducing GHG emissions and eager to learn from their peers in other countries. They share an entrepreneurial spirit and the desire to shape policy at EU level (cf. Kassim 2005). The Covenant of Mayors is an excellent example of the transformation of the EU multilevel climate governance system and the emergence of new forms of vertical climate governance that create direct relations between the most active cities and the EU.

HORIZONTAL CLIMATE GOVERNANCE IN THE EUROPEAN UNION

In recent years horizontal climate governance has gained in importance and provides additional opportunities for cities. Three types of horizontal climate governance can be found in the European Union. First, the most traditional type of horizontal governance is city twinning which has a tradition that is much older than the European Union because it developed as early as after World War I, but more particularly after World War

II. Sister city arrangements are long-term relationships of a rather general nature but can be used as a basis for more elaborated forms of cooperation.

Second cities may become involved in project networks, in particular in projects funded by the EU. Such project networks usually include more than two cities, but the number of cities is still limited in order to facilitate knowledge transfer and learning among the participating cities. Project networks with cities from different countries create transnational spaces beyond nation states. Such networks can include sister cities, but these networks are most often limited to the duration of a specific project. Thus such networks are more volatile and may disappear after the end of the project. They may however, also be used to apply for a new project with the aim of extending the cooperative relationship and finding joint solutions for common problems.

City twinning and project networks can serve as an excellent basis for the establishment of transnational city networks, the third type of horizontal climate governance. As discussed in the previous section, transnational city networks such as Energy Cities are a combination of vertical and horizontal climate governance. Apart from lobbying, city networks are created to facilitate the exchange of experience and foster the transfer of knowledge among member cities (Kern 2001; Kern and Bulkeley 2009; Román 2010; Lee and van de Meene 2012). Membership in a transnational city network such as Energy Cities provides an excellent basis for finding city partners, if this is required for a project application. The fact that three transnational city networks in the area of climate and energy policy had emerged by the early 1990s can be explained, at least partially, by the international and transnational nature of the European Union itself. In other parts of the world, transnational city networks were established as well but, with respect to the institutional density of this type of arrangement, the development in Europe is rather unique.

These three types of horizontal climate governance show that cities learn not only from their own experience but also from other cities. Sister city agreements, cross-border project networks and transnational city networks support the transfer of knowledge, mutual learning and the joint development of innovative solutions. Knowledge transfer and learning depend on three elements: (1) the transferability of new approaches across cities; (2) the existence of transnationally oriented policy entrepreneurs, such as an internationally engaged deputy mayor; and (3) appropriate institutional arrangements that facilitate knowledge transfer. Successful approaches may change considerably during the transfer process because they need to be adapted to a new institutional context.

Contextual factors eventually determine how policy models are utilized (Radaelli 2005; Theobald and Kern 2011). Moreover, the diffusion of policy approaches is not always a self-sustaining process, even if successful models and sufficient capacities exist. Nonadoption may be a rational strategy, in particular when powerful and dominant players oppose such policy changes.

CONCLUSIONS

The discussion has shown that urban climate governance in the EU multilevel system can take three different forms. Hierarchical climate governance restricts cities to implementing EU and national climate change policy, and is somewhat exceptional in EU climate governance because in most countries there is no direct relationship between EU or national GHG reduction goals and local politics. Vertical climate governance means that cities have started to influence EU climate change policy directly and have established direct links with EU institutions, in particular through the Covenant of Mayors. Horizontal climate governance works through bilateral and multilateral city networks that aim to facilitate knowledge transfer and learning. The establishment of new institutional arrangements, in particular transnational city networks, such as the Climate Alliance, combine vertical and horizontal forms of climate governance in a productive way because they allow cities to become active at the EU level and they also facilitate knowledge transfer and learning.

Although cities such as Stockholm, Copenhagen, Amsterdam and Munich have become internationally acclaimed leaders, other cities and towns in Sweden, Denmark, the Netherlands and Germany are not. Even in the leading countries we find a split between the pioneering cities and cities which clearly lag behind. Traditional policy instruments such as the Dutch BANS agreement and subsidy scheme cannot solve this problem. Existing differences may become even more pronounced because forerunners may apply for funds and run successful programs, leaving the less innovative cities behind.

European Union multilevel climate governance requires integrated approaches to prevent and correct the split between the pioneering cities and the laggards. Institutional innovations such as the establishment of the Covenant of Mayors are attempts to cope with the disparities of local climate governance across Europe. Such institutional innovations provide tools for the vertical and horizontal integration of EU climate governance. However such arrangements face a dual challenge because they are confronted with not only the complexity of the climate change problem, the uncertainty of its impacts and the ambiguity of decision making in this policy area, but also with the disparities of local climate governance in Europe.

While hierarchical climate governance neglects bottom-up initiatives, vertical climate governance combines local and EU initiatives. The options for cities to gain access to EU climate governance have improved considerably since some pioneering cities started their initiatives about 20 years ago. Today the European Union provides an opportunity structure, which allows cities to gain access to different EU institutions, in particular the EU Commission. This development has been supported by the emergence and stabilization of three transnational city networks in the area of climate and energy governance. These networks are a combination of vertical and horizontal climate governance because their main goals are the representation of their members' interests at EU level and the facilitation of knowledge transfer and learning across European cities. Institutional innovations such as the Covenant of Majors are a new step in this development because they transform traditional forms of hierarchical climate governance, which restricts local authorities to the implementation of EU and national legislation, by supplementing it with a new type of vertical climate governance that bypasses the Member States.

Transnational networking in the area of climate change policy has been the privilege of the pioneering cities that have the capacities to initiate and maintain such activities. Thus, transnational city networks can be regarded as networks of pioneers for pioneers (Kern and Bulkeley, 2009). As pioneering cities can be found in all Member States, a cluster of truly Europeanized cities which take part in many European and international campaigns has emerged. The dynamic development of the Covenant of Mayors shows however, that 'second tier' cities can also become involved in European campaigns when they are supported by provinces or city associations. Moreover, the recent development is also characterized by an increase of networking at (micro-) regional scale, in particular in densely populated city regions. Transnational city pioneers may become models and influence the climate change policy of neighbouring cities. In addition, many cities have realized that climate change policy should not stop at city borders because many problems related to climate mitigation and adaptation require new forms of cross-border governance at regional scale.

The EU multilevel system provides new opportunities for cities to boost their local climate initiatives by joining European networks and campaigns. Whether they use these opportunities depends mainly on a city's commitment to reduce GHG emissions, develop a low carbon infrastructure, adapt to climate change impacts and help its citizens to adjust to the changing situation. Pioneering cities like Stockholm, Copenhagen, Amsterdam and Munich are far ahead, can make a difference and may even serve as models at a regional, national, European and international level.

NOTE

1. With around 1400 full members in 20 European countries, the *Climate Alliance*, whose headquarters are located in Frankfurt am Main, is by far the largest of these networks. *Cities for Climate Protection* (CCP) has around 170 member cities in 19 European countries. It is an ICLEI initiative – a global transnational city network whose European headquarters are in Freiburg (Germany). *Energy Cities* has close to 200 individual members in 31 European countries and its headquarters in Besançon (France) (all data for September 2012).

REFERENCES

Allen, D. (2005), 'Cohesion and the Structural Funds,' in H. Wallace, W. Wallace and M. Pollack (eds), *Policy-Making in the European Union*, Oxford: Oxford University Press, pp. 213–41.

Ausschuss der Regionen (2010), *Weissbuch des Ausschusses der Regionen zur Multi-Level-Governance,* Partnerschaft für den Aufbau Europas, Brüssel: Ausschuss der Regionen.

Bache, I. (2008), *Europeanization and Multilevel Governance*, Lanham: Rowman & Littlefield.

Bache, I. and M. Flinders (eds) (2004), *Multi-level Governance*, Oxford: Oxford University Press.

Börzel, T. and T. Risse (2007), 'Europeanization: the Domestic Impact of European Union Politics,' in K. Jørgensen, M. Pollack and B. Rosamond (eds), *Handbook of European Union Politics*, London: Sage.

Van Bommel, S. and W. Kuindersma (2008), *Policy Integration, Coherence and Governance in Dutch Climate Policy*, Wageningen: Alterra.

Bourne, A. (2007), 'Regional Europe,' in M. Cini, (ed.), *European Union Politics*, 2nd ed., Oxford: Oxford University Press, pp. 287–303.

Bulkeley, H. and K. Kern (2006), 'Local Government and the Governing of Climate Change in Germany and the UK,' *Urban Studies* 43(12), 2237–59.

Bulmer, S., D. Dolowitz, P. Humphreys and S. Padgett (2007), *Policy Transfer in European Union Governance, Regulating the Utilities*, London/New York: Routledge.

Carbon Disclosure Project (2012), *Seven Climate Change Lessons from the Cities of Europe*, CDP Cities 2012.

Citi, M. and M. Rhodes (2006), 'The Modes of Governance in the European Union: A Critical Survey and Analysis,' in K. Jørgensen, M. Pollack and B. Rosamond (eds), *Handbook of European Union Politics*, London: Sage.
Delmas, M. and O. Young (eds) (2009), *Governance for the Environment. New Perspectives*, Cambridge: Cambridge University Press.
Den Exter, R. (2012), *Climate Policy in Dutch Municipalities: Organization, Policy, Implementation and Performance*, MSc. Thesis, Wageningen University.
Deutsches Institut für Urbanistik (ed.) (2011), *Klimaschutz in Kommunen. Praxisleitfaden*, Berlin: Deutsches Institut für Urbanistik.
Enderlein, H., S. Walti and M. Zurn (eds) (2010), *Handbook on Multi-level Governance*, Cheltenham: Elgar.
European Commission (2001), *European Governance: A White Paper,* Luxembourg: European Communities.
Evans, M. (ed.) (2004), *Policy Transfer in Global Perspective*, Aldershot/ Burlington: Ashgate.
Fudge, S. and M. Peters (2009), 'Motivating Carbon Reduction in the UK: The Role of Local Government as an Agent of Social Change,' *Journal of Integrative Environmental Sciences* 6(2), 103–20.
Goldsmith, M. (2003), 'Variable Geometry, Multi-level Governance: European Integration and Subnational Government in the New Millennium,' in K. Featherstone and C. Radaelli (eds), *The Politics of Europeanization*, Oxford: Oxford University Press, pp. 112–33.
Gupta, J. (2007), 'The Multi-level Governance Challenge of Climate Change,' *Environmental Sciences* 4(3), 131–7.
Gupta, J., R. Lasage and T. Stam (2007), 'National Efforts to Enhance Local Climate Policy in the Netherlands,' *Environmental Sciences* 4(3), 171–82.
Gustavsson, E., I. Elander and M. Lundmark (2009), 'Multilevel Governance, Networking Cities and the Geography of Climate-Change Mitigation: Two Swedish Examples,' *Environment and Planning C: Government and Policy* 27, 59–74.
Heinelt, H. and S. Niederhafner (2008), 'Cities and Organized Interest Intermediation in the EU Multi-level System,' *European Urban and Regional Studies* 15(2), 173–87.
Hooghe, L. and G. Marks (2003), 'Unraveling the Central State? Types of Multi-level Governance,' *American Political Science Review* 97(2), 233–43.
John, P. (2000), 'The Europeanisation of Sub-national Governance,' *Urban Studies* 37(5-6), 877–94.
John, P. and C. Copus (2011), 'The United Kingdom: Is There Really an Anglo Model?' in J. Loughlin, F. Hendriks and A. Lindström (eds), *The Oxford Handbook of Local and Regional Democracy in Europe*, Oxford: Oxford University Press.
Karch, A. (2007), *Democratic Laboratories. Policy Diffusion among the American States*, Ann Arbor: University of Michigan Press.
Kassim, H. (2005), 'The Europeanization of Member state Institutions,' in S. Bulmer, and C. Lequesne (eds), *The Member States of the European Union*, Oxford: Oxford University Press, pp. 285–316.

Kern, K. (2000), *Die Diffusion von Politikinnovationen. Umweltpolitische Innovationen im Mehrebenensystem der USA*, Opladen: Leske and Budrich.

Kern, K. (2001), 'Transnationale Städtenetzwerke in Europa,' in E. Schröter (ed.), *Empirische Policy- und Verwaltungsforschung. Lokale, nationale und internationale Perspektiven*, Opladen: Leske and Budrich, pp. 95–116.

Kern, K. and H. Bulkeley (2009), 'Cities, Europeanization and Multi-level Governance: Governing Climate Change Through Transnational Municipal Networks,' *Journal of Common Market Studies* 47(2), 309–32.

Kern, K., C. Koll and M. Schophaus (2007), 'The Diffusion of Local Agenda 21 in Germany: Comparing the German Federal States,' *Environmental Politics* 16(4), 604–24.

Kooiman, J. (2003), *Governing as Governance*, London: Sage.

Kooiman, J. and S. Jentoft (2009), 'Meta-Governance: Values, Norms and Principles and the Making of Hard Choices,' *Public Administration* 87, 818–36.

Lee, T. and S. van de Meene (2012), 'Who Teaches and Who Learns? Policy Learning through the C40 Cities Climate Network,' *Policy Sciences* 45(3), 199–220.

Marsh, D. and J.C. Sharman (2009), 'Policy Diffusion and Policy Transfer', *Policy Studies* 30(3), 269–88.

McCann, E. and K. Ward (2012), 'Assembling Urbanism: Following Policies and 'Studying Through' the Sites and Situations of Policy Making,' *Environment and Planning*, A 44, 42–51.

Monni, S. and F. Raes (2008), 'Multilevel Climate Policy: The Case of the European Union, Finland and Helsinki,' *Environmental Science and Policy* 11, 743–55.

Niederhafner, S. (2008), Europäische Städte im Mehrebenensystem der EU. Eine vergleichende Untersuchung der Einflussnahme kommunaler Akteure auf europäischer Ebene anhand der beiden pan-europäischen Kommunalverbände Eurocities and RGRE*, Dissertation TU Darmstadt.*

NOVEM (2003), *The Local Approach to Climate Policy in the Netherlands*, Sittard: Utrecht.

Organization for Economic Cooperation and Development (OECD) (2010), *Cities and Climate Change*, OECD Publishing, Paris.

Ostrom, E. (2010), 'Polycentric Systems for Coping with Collective Action and Global Environmental Change,' *Global Environmental Change* 20, 550–57.

Pierre, J. and B.G. Peters (2000), *Governance, Politics and the State*, Houndmills, UK: Macmillan.

Radaelli, C. (2005), 'Diffusion without Convergence: How Political Context Shapes the Adoption of Regulatory Impact Assessment,' *Journal of European Public Policy* 12, 924–43.

Rechlin, S. (2004), Die deutschen Kommunen im Mehrebenensystem der Europäischen Union – Betroffene Objekte oder active Subjekte?, WZB Discussion Paper SP IV 2004-101, Berlin.

Román, M. (2010), 'Governing from the Middle: The C40 Cities Leadership Group,' *Corporate Governance* 10, 73–84.

Rosamond, B. (2007), 'New Theories of European Integration,' in M. Cini (ed.), *European Union Politics*, 2nd ed., Oxford: Oxford University Press, pp. 117–36.
Rose, R. (2005), *Learning from Comparative Public Policy,* London/New York: Routledge.
Rosenau, J. (2003), *Distant Proximities. Dynamics beyond Globalization*, Princeton: Princeton University Press.
Sbragia, A. (2004), 'Key Policies,' in E. Bomberg and A. Stubb (eds), *The European Union: How Does it Work?*, Oxford: Oxford University Press, pp. 111–35.
Schulze, C. (2003), 'Cities and EU Governance: Policy-takers or Policy-makers?' *Regional and Federal Studies* 13(1), 121–47.
Smith, A. (2007), 'Emerging in Between: The Multi-level Governance of Renewable Energy in English Regions,' *Energy Policy* 35, 6266–80.
Theobald, H. and K. Kern (2011), 'The Introduction of Long-term Care Policy Schemes: Policy Development, Policy Transfer and Policy Change,' *Policy & Politics* 39(3), 325–42.
VROM (2007), *Vervolgsubsidieregeling BANS Klimaatconvenant 2007*, Ministry of Housing, Spatial Planning and the Environment, The Hague.
Westerhoff, L., E. Carina, H. Keskitalo and S. Juhola (2011), 'Capacities across Scales: Local to National Adaptation Policy in Four Countries,' *Climate Policy* 11, 1071–85.

7. Bottom-up versus top-down: the evolving American climate policy odyssey

Barry G. Rabe

Three of America's most prominent political leaders gathered in May 2009 on the White House lawn to announce an intergovernmental pact on climate change. President Barack Obama declared a new agreement on vehicle fuel efficiency and tailpipe emissions of carbon dioxide, establishing stringent new national standards intended to reduce American greenhouse gas (GHG) emissions from the transportation sector. California Governor Arnold Schwarzenegger beamed his consent, as this step represented the formal federal government embrace of legislation passed seven years earlier in Sacramento and later backed by 14 other states. Michigan Governor Jennifer Granholm was also on hand to endorse the deal, even though her state had earlier opposed the efforts of other states to go into the federal courts to seek such a decision.

This tableau underscores the fact that climate change policy has increasingly – and unexpectedly – assumed an intergovernmental dimension. This is true in formal federations such as the United States as well as for other governments around the world that foster some degree of cross-level governance. Of course, this challenges much conventional analysis, as climate change was long framed as anything other than an intergovernmental issue. Since its initial arrival on both the international and American domestic agendas in the 1970s and 1980s, climate change has been commonly presumed to entail a dominant role for major national powers that would cobble together a multinational pact to be implemented by an international regime. Such a view has been evident in each major milestone of international policy development, including the 1992 United Nations Framework Convention on Climate Change, the 1997 Kyoto Protocol and all subsequent international conclaves such as annual meetings of the Conference of Parties.

The United States has produced more GHGs than any other nation over recent decades; it continues to generate more than one fifth of global totals despite being passed in 2007 by China for world leadership in annual emissions. American per capita emissions continue to rank among the very highest in the world. Given this outsized role, no international attempt to reduce GHG emissions and reduce the threat of climate change can realistically advance far without active American engagement. At the same time, unilateral American efforts, regardless of their scope, cannot resolve this issue.

This chapter explores the intergovernmental complexities of climate policy through an analysis of the evolving American case. It will examine three distinct periods of American climate policy, each featuring differing federal and state roles and relationships. This will begin with a review of the extended period from 1975 and 1997 when both the federal government and many states took modest, often symbolic steps to define the issue and began to consider policy options. We will then consider the transition into an extended period of 'state domination' of climate policy development from roughly 1998 through 2007, as federal disengagement opened a clear playing field for state policy experimentation. This will be followed by the current era of 'contested federalism,' whereby the federal government was prompted by a US Supreme Court decision and a major electoral transition to consider a much expanded role amid established state policy commitments. The discussion of contested federalism will include review of alternative forms of divided or shared federal and state authority that began to be explored during this period. This includes such options as pre-emption that would entail a dominant federal role or a more collaborative form of federalism that would blend authority and attempt to accentuate the respective strengths of both levels. The chapter concludes with a review of some of the challenges and opportunities posed by this extended intergovernmental evolution for the future of American involvement in climate policy, as well as a review of the varied set of policy outcomes that have emerged thus far in the period of contested federalism. It will note that the possibility of a federal takeover of this arena seems unlikely to occur in the near future, leaving a complex mixture of intergovernmental responsibilities in the coming years and also the possibility of a more state-centric intergovernmental system than seemed plausible as recently as 2009.

CLIMATE CHANGE, AMERICAN FEDERALISM STYLE

Lacking precise intergovernmental dividing lines, scholars have searched for criteria to explain why the policy approaches of various levels of American government shift over time and frequently diverge. Many have defined and examined patterns whereby states (either a single state or a subset of them) tend to take the lead in policy innovation, sometimes fostering considerable subnational diffusion of policy that may (or may not) trigger some form of federal response (Volden et al. 2008). States are often found taking early action to gain competitive advantage over their neighbours, most commonly through attempts to accelerate or diversify economic development. Such steps are often thought to coincide with state aversion to those governmental functions that could impose heavy costs on a state and its taxpayers such as regulatory or redistributive programs (Peterson 1995). Questions of federal engagement often follow, whether to build on early state efforts through policy learning, erase them in the name of national uniformity, or jump in amid a seeming absence of state involvement (Posner 2010).

Application of this work to the arena of environmental protection regularly raises issues of negative externalities and whether states attempt to advance their own economic well-being by generating environmental contaminants that drift or flow to other jurisdictions where they cause ecological or human health damage (Lowry 1997). This has fostered considerable analysis of strategies to either reconcile crossjurisdictional conflicts (such as interstate litigation for alleged damages or the creation of interstate compacts to unite neighbouring states on common environmental concerns) or sort out federal and state responsibilities to minimize cross-state contaminant transfer (such as federal pre-emption of a policy area). Externalities are rarely easy to measure with precision, however, and periodic efforts to establish clear dividing lines between what constitutes a federal as opposed to a state area of responsibility have proven elusive at best.

Analysis that focused on intergovernmental relations was initially dismissed by scholars as irrelevant to climate change, given the predominant thinking that theories of international relations and regime formation were far more applicable. As we shall see, climate change policy options pose classic questions of economic development versus cost imposition. Just as no two states are likely to face the same risks from accelerating climate change, no two will likely frame such policy options in identical fashion or have a comparable capacity to implement policy.

These factors suggest that differing scenarios may emerge as units within a federal United States consider climate policy options. As indicated in Table 7.1, both federal and state government involvement can run along a high-to-low continuum, with possibilities of significant changes over time. Each possible combination presents a different intergovernmental dynamic, with the United States appearing to have moved from one domain to another in recent years. In the case of low federal *and* state involvement (Cell 4), there is scant intergovernmental engagement because little or no policy is being enacted at either level. This is reflected in the extended period of largely symbolic policy that prevailed between 1975 and 1997. In the case of high federal and low state involvement (Cell 3), the federal government has established a dominant relationship. There may be intergovernmental tension over how those federal policies are implemented but the central government role is firmly established. This can be described as 'federal domination.' This approach seemed increasingly plausible politically in the late 2000s but efforts to expand the federal role legislatively and constrain state authority stalled in 2010 and appeared unlikely to emerge in succeeding years. This left a series of uncertainties about future directions, including questions about the sustainability of state policy commitment and possible federal engagement via executive branch actions that might be vulnerable to future political shifts.

In the case of low federal and high state involvement (Cell 2), states will likely be exploring just how far they can go given the confines of a federal system and possible backlash from either federal or rival state authorities. This can be described as 'state domination' and reflects the decade-long period that began in 1998. In the case of high federal and high state involvement (Cell 1), both levels are actively involved and may either collide, compete or cooperate with one another. This can be characterized as 'contested federalism' and is reflected in a number of more recent developments. As we shall discuss further, the United States appears to have spent much of the past decade in Cell 2. It has subsequently shifted towards the contested federalism of Cell 1.

We shall explore the evolution of American climate change policy in recent decades using this framework, devoting particular attention to the dynamics within each cell at a given time period as well as factors that cause movement across cells at key intervals.

SYMBOLIC POLICY: 1975 TO 1997

Climate change did not arrive on the doorsteps of federal and state capitals in the late 1990s, when deliberations leading to the ill-fated Kyoto Protocol took place. Instead, it surfaced much earlier at both

Table 7.1 Intergovernmental involvement in American climate change policy, 1975–2013

	High federal involvement	Low federal involvement
High state involvement	1 Contested Federalism, 2008–2013	2 State Domination, 1998–2007
Low state involvement	3 Federal Domination (None)	4 Symbolic Policy, 1975–1997

levels, focused primarily on scientific interpretation but also generating very modest policy steps. At the federal level, the executive branch first turned to this issue in the Domestic Policy Council during the Gerald Ford presidency, given early evidence of what is now characterized as climate change. This overlapped with a dramatic and disruptive spike in oil prices, which heightened federal attention on energy supply and policy alternatives. Congressional hearings on climate change began in 1975, followed by a slow but steady pace of additional hearings until President Ronald Reagan signed the 1987 Global Climate Protection Act into law, establishing an interagency task force on this issue. One year later, Republican Governors in California and New Jersey signed the first of what would become a steady stream of state laws intended to address climate change.

None of these early steps, however, approached any reasonable threshold of what constitutes a serious policy initiative. A common theme was early problem recognition and very modest initial steps to develop a long-term governance strategy. No subsequent federal legislation or executive orders followed during the balance of this period and most state policy development during the 1975–1997 time span involved nonbinding policy recommendations, voluntary reduction programs or small scale renewable energy or energy efficiency programs. Consequently, this extended period allowed scientists, journalists, environmental groups and state and federal policy entrepreneurs to broaden their study of climate change and begin to formulate policy options. Hence climate change reached federal and some state government agendas but remained a low level concern. It did not generate serious engagement by either level, much less any form of intergovernmental tension or competition.

STATE DOMINATION: 1998 TO 2007

The American intergovernmental equilibrium on climate change was shattered towards the end of the 1990s, once the United States negotiated and then signed the Kyoto Protocol in 1998. If ratified by the Senate, Kyoto implementation would indeed have involved the federal government in a major way. But Kyoto had no real prospect of Senate ratification and the Bush Administration's 2001 decision to withdraw formally only made official the reality that the treaty had no political future in the United States. This recognition, alongside the continued inability of Congress to advance serious climate policy, created an open intergovernmental field for state government engagement, analogous to other policy domains which have been vacated by the federal government without precluding state involvement and thereby creating the possibility of ongoing state government domination (Nathan 2008).

Less than a decade after the Kyoto rejection by the federal government, a large and ever expanding body of American states were laying claim to national and, in some instances, global leadership on climate change through unilateral policy development. This entailed experimentation in one or (usually) more states with nearly every policy tool imaginable that could achieve some GHG reduction. Much of this state policy development was locally developed and yet followed two well established patterns for expansion within a federal system such as the United States: diffusion and interstate compacts. First, a number of climate policies followed a classic pattern of diffusion, whereby a policy idea begins in one state but then is essentially replicated elsewhere. For example, renewable electricity mandates represent one way to facilitate a transition away from fossil-based electricity, operational in 29 states in 2013. These so called 'renewable portfolio standards' (RPS) are most heavily concentrated in the Northeast, Pacific West and Southwest but are in place in one or more states in every region of the nation. No two states have identical RPSs but they generally follow a similar structure. Many states have established statutory targets to reach between 15 and 25 per cent of their electricity from renewable sources over the next decade, which would translate into significant GHG reductions from that sector if implemented.

Second, the US Constitution allows state governments to establish regional policies across borders through interstate compacts ratified by Congress. States have also experimented with the simultaneous development of comparable policies that can feature collaborative implementation, though there are constitutional uncertainties concerning how far these can go (Farber 2008; Kysar and Meyler 2008). Prominent examples of the latter approach are evident in three multistate regions established through interstate agreement to operate carbon cap-and-trade programs. At their peak, these collaborations involved 23 states, including ten from the Northeast (Regional Greenhouse Gas Initiative or RGGI), seven from the West (Western Climate Initiative or WCI) and six from the Great Lakes (Midwestern Greenhouse Gas Reduction Accord or MGGRA). The latter two secured formal participation commitments from four Canadian provinces: British Columbia, Manitoba, Ontario and Quebec. Each regional zone involves a memorandum of understanding between these jurisdictions that is put into operation through state or provincial legislation or regulation. The Regional Greenhouse Gas Initiative is furthest advanced, having begun carbon auction allowances for the electricity sector among its members in 2008 and generated more than $1 billion for member states during its first four years of operation. Both WCI and MGGRA pledged to reduce regional carbon emissions from a much wider range of sources to 15 per cent below 2005 levels by 2020. However, the WCI lost all state members except California by 2012 and the MGGRA imploded amid political leadership changes in most members by 2011. California planned to launch its auctioning system in 2013 and hoped to eventually lure other states into partnership. These state initiatives in the absence of federal involvement indicate the substantial incentives that states (and their leading policy makers) might have to take significant climate policy steps, as well as the challenge of sustaining such initial commitments into implementation.

Strategic Economic Advantage. State governments have long been noted for their capacity to use diverse policy tools to promote internal economic development. On the one hand unilateral efforts to reduce GHGs might indeed be seen as detrimental to development, most likely through increasing energy costs. But a small army of state governors and other state political leaders have promoted climate policies in recent years, guided by a rationale based substantially on economic development grounds. Consequently, states might pursue renewable energy mandates or carbon cap-and-trade programs in part for climate and environmental concerns but also (and perhaps more significantly) for perceived economic advantages. This could entail development of 'home grown' sources of electricity, such as wind and biomass, as well as technologies and skills that could be in considerable demand in a carbon-constrained world. In turn, states carefully downplay any potential costs that these policies might impose upon citizens. They have proven averse to policy options such as carbon or energy taxes that might deter fossil fuel use but also alienate citizens by directly increasing their energy costs (Rabe 2008).

Just as states have raced to the top in other areas, including development of nanotechnology and stem cell medical technology, an ever growing number have jumped into the race for climate friendly economic development in recent years. The Great Recession provided added impetus to promote alternative energy policy as a state economic development option. As California Air Resources Board Chair, Mary Nichols, said in 2013, as her state began to move into full climate policy implementation, 'By valuing carbon, we will foster the most competitive, productive businesses in the world. We will not only lead the world in combating climate change, but also build lasting resiliency into California's economy' (Nichols 2013).

Such framing opened state receptivity to a range of ways to promote these resources and skills, including regulatory mandates and financial incentive programs. Some states have also taken policy steps designed to protect against possible economic risk, in some cases betting against conventional, high carbon technologies in anticipation of ever expanding concerns over their environmental imprint. In recent years, a number of states have used their regulatory powers to thwart proposed coal burning power plants, even though these produce substantial energy from an abundant source at relatively low cost. There have been a number of reasons behind this new strategy, including the expanded availability of natural gas through hydraulic fracking techniques, but consideration of economic ramifications has been a big part of the equation.

Still another state economic development incentive may relate to policy opportunities that, in effect, will shift most of the compliance costs to other jurisdictions. California's alliance with other states pursuing vehicle emissions reductions can be considered through this lens, as none of these jurisdictions host large vehicle manufacturing sectors that might be jeopardized through aggressive transition towards lower emission vehicles. The citizens of these states might ultimately pay more for new vehicles but no more than their neighbours around the nation, and they might derive considerable savings if these policies fostered substantial reductions in per mile transportation costs. In turn, some of the proponent states were actively developing next-generation vehicle technology within their boundaries that might receive a boost through a regulatory burden imposed on conventional vehicles generally manufactured elsewhere. Under such a cost transfer strategy, any environmental benefits linked to reduced GHGs or other emissions would only be an added benefit. States might further angle for benefit maximization through early policy steps that would position them to extract maximum advantage under any future federal regulatory regime, building on environmental policy precedents whereby early state action has led to either expansive monetary rewards through grants or formal credit for early emissions reductions (Rabe 2008).

Strategic Political Advantage. States might also choose to explore unilateral policy steps to reduce GHG emissions in order to burnish their reputations for innovation on a larger stage, perhaps national or even international. At the same time, individual political credit for championing such policy innovation might be considerable. This phenomenon has been most evident among governors, perhaps best reflected in California, where Schwarzenegger's 2006 electoral fortunes were clearly enhanced by his collaboration with Democratic legislators to produce ambitious climate legislation just before November's voting. This was illustrated in dramatic bill signing ceremonies in which California laid claim to global leadership on climate change and Schwarzenegger's evolution from film celebrity to environmental policy superhero appeared complete. The 2006 experience was followed by the governor's decision to host three 'global climate summits' between 2008 and 2010 in Beverly Hills.

Not every policy champion is as visible as Arnold Schwarzenegger, but a diverse array of both elected and appointed officials in many states have taken active roles as policy entrepreneurs. This entails nurturing a policy idea, building a supportive political coalition and guiding this effort to enactment through a 'steering network' (Bomberg 2009). A primary reward is the making of 'good policy' that might improve the environment and the climate. But studies of policy entrepreneurship also suggest that there are multiple incentives for officials to take such steps, including increased prominence and possible career advancement (Teodoro 2011). In state politics this might entail added visibility to increase prospects for seeking another office, whether due to ambition to climb the political ladder or the spectre of unemployment linked to formal term limits in many state offices.

Such entrepreneurial activity need not be confined to elected officials. Many states have proven fertile settings for individuals who serve in various departments and agencies to perform similar roles, often very close to the particular details of a policy and responsible for luring elected officials into playing a supportive role (Rabe 2004). Such individuals may be unit leaders appointed by a governor or career staff. They may also be given an entrepreneurial opportunity in state government through outside support, such as foundations that provide grants to consultants who work with states by advising governors, facilitating statewide action plans or providing technical assistance. Advancing climate policy innovation may literally be the bureaucratic 'role of a career,' reflecting a unique blend of expertise and opportunities for advancement. A number of the appointed state agency officials who were instrumental in RGGI development, for example, received significant new career opportunities in the Obama Administration.

Filling the Intergovernmental Void. This blend of economic and political incentives encouraged an unexpectedly large set of states to explore active engagement in climate policy. States' substantial constitutional authority made development of policy to reduce GHG emissions a fairly incremental extension of existing authority rather than a radical new departure. Under this context, economic and political advantages served to accelerate policy development that 'made sense' for a given jurisdiction. In turn, states attempted to fill the substantial intergovernmental void created when the federal government essentially abandoned the playing field of climate change policy in the late 1990s. This gave enormous latitude to states to do nothing, pursue a few symbolic steps, enact one or two significant policies, or pursue a far-reaching approach that might position them for regional and national leadership and even global visibility. Clusters of states fall into each of these categories, all emerging from a period of state domination of American climate policy development that began in the late 2000s to begin to consider a potentially large expansion of the federal role.

CONTESTED FEDERALISM: 2008 TO 2010

The first significant indication that the American intergovernmental balance on climate policy might be changing occurred when the US Supreme Court performed the role of intergovernmental umpire. Massachusetts and 12 allied states contended that the federal government's refusal to designate carbon dioxide as an air pollutant under the 1990 Clean Air Act Amendments placed them in danger of such risks as sea level rise due to climate change. Ten other states took the opposite position, backing the Bush Administration's claim that the federal government lacked statutory authority and that states had no business being in court on such a matter. In 2007 a five-to-four majority of justices ruled in favour of forcing the Environmental Protection Agency (EPA) to revisit its refusal to designate carbon dioxide as an air pollutant. This decision represented a federal court response to state pressure to compel federal executive branch action, with potentially far reaching intergovernmental consequences.

With this decision, as well as the 2008 national election that appeared to open a window to expanded federal involvement, the era of state domination of American climate policy appeared to be effectively over. It was supplanted by a new era of contested federalism reflected in continued high involvement by states but also increasingly high involvement by the federal government (Cell 1). In some areas the details of an expanding federal role have become clear, particularly through such Obama Administration policies as its intergovernmental agreement to reduce transportation-based emissions, which was later expanded through 2025. Some Congressional consensus appeared to be emerging through June 2009 with House approval of far-reaching climate legislation that included an expansive cap-and-trade program with substantial consequences for existing state policies. This form of legislation, however, failed to gain support in the Senate and appeared highly unlikely to be revisited in succeeding Congresses. Consequently, federal climate policy development remained concentrated within the executive branch. The Obama Administration in 2009 accepted the Supreme Court prod and included carbon dioxide under the federal definition of an air pollutant. It has since followed this reframing through EPA efforts to launch a permit process to reduce GHG emissions from large sources under the auspices of the Clean Air Act, albeit heavily reliant on states for interpretation and implementation. These federal efforts, however, have begun to face potential challenges through litigation as well as possible Congressional action. Consequently, this creates an uncertain context for federal policy and thereby increases the likelihood of retaining a state-centric system in coming years.

The period of contested federalism has emerged in a remarkable intergovernmental context, one where approximately half of the states are implementing multiple climate policies, but the remaining states are less involved. So any federal movement into this arena could have significant consequences for various states and regions. As the 111th Congress began, did the federal government see climate change as an opportunity to take full control of this arena or somehow share authority with states? As climate policy proposals emerged from the President and Congress, did these approaches to federal policy give some form of credit to those states that had taken early action and achieved emission growth levels below the national average? Or did they concentrate incentives on those states with little policy development and high emissions trajectories? Prior experience in American federalism suggested that there might not be one uniform strategy to guide this issue given the many dimensions and sectors with some link to climate change. A range of possible options received increasing scrutiny from Congress, the Obama Administration and the states, each with distinct intergovernmental implications. This shifting dynamic obviously influences future state strategic decisions, moving from a playing field in which they had no intergovernmental rival to one in which they could be constrained or even eviscerated at any moment. Subsequent sections will consider alternative ways in which contested federalism began playing out based on the experience of the past few years.

State Positioning. In anticipation of an expanded federal role, states began to position themselves to influence federal policy, both through associations representing all 50 states as well as individually. Consequently, one could begin to consider states, both collectively and individually, as strategic actors engaged in intergovernmental lobbying in search of most favoured status as the federal government moved onto terrain that states had long dominated. In some instances this entailed state alliances with other entities, including industries and environmental advocacy groups.

Organizations that represent the views of all states must of course contend with differences among their membership but generally find consensus positions that allow them to take fairly uniform stands in attempting to influence federal policy. Virtually all of these state-based entities took a fairly similar stance on possible expansion of the federal role in climate change, reflected in position papers, policy briefs, public workshops and formal testimony. They generally tended to endorse intergovernmental strategies that would protect existing state policies and allow for continued state innovation. They also sought to extract as much rent as possible, in the form of grants and other financial support, from the federal government to cover implementation costs, further promote their most promising renewable energy sources and underwrite efforts to 'adapt' to changing climates. Among those associations that represent elected state officials, for example, the National Governors Association and National Conference of State Legislatures (NCSL) took generally similar positions. Those organizations that represent state agencies with a common function, such as the Environmental Council of the States (environmental protection agencies), the National Association of Regulatory Utility Commissioners (electricity regulatory boards), the National Association of State Energy Officials (energy departments) and the National Association of Clean Air Agencies (state and local air quality units), also tended to take similar stances, though tailored to their particular area. In short, these groups sought to protect state interests under contested federalism, whether giving states latitude to sustain existing policies or to take additional steps in the future.

At the same time, some individual states began to break ranks and sought preferential treatment for themselves under any expanded federal policy. Those states that sustained relatively low patterns of emissions growth and high levels of policy development sought special recognition in any federal cap-and-trade policy. This included supplemental credits for reductions that took place before a federal program began, as had occurred in the 1990s during the creation of an emissions trading system for sulfur dioxide. States increasingly began to pursue a pair of strategies simultaneously in intergovernmental negotiations, taking both collective stands as well as ones tailored to their particular advantage.

In response, federal officials considering policy engagement had to weigh how seriously to consider these collective and individual state claims, given the different ways in which state and local interests are represented in the two chambers of Congress and the significant regional differences in energy production and consumption that are so central to climate policy. In the House, California (with the lowest per capita rate of carbon emissions of any state) has more than 50 times the voting power of Wyoming (with the highest per capita rate of carbon emissions of any state), reflecting their respective populations. But in the Senate, the two states have identical voting power. These realities have long made the intersection of energy and environmental protection among the most contentious in American politics, given varied degrees of economic dependence between states on the extraction of fossil fuels and their use in meeting core energy needs (Lowry 2008). Any federal policy would have to run this political gauntlet successfully, while also weighing the positions of the states against all other organized interests. Indeed, each interest is likely to prefer its own balance between federal and state authority in any emerging climate policy. This raises the possibility of very distinct policy alternatives, each of which would tilt the intergovernmental system in very different ways.

Total Pre-emption: Federal Encroachment. One option for federal governments that move into a state dominated policy arena is complete takeover, eviscerating any state role and imposing federal domination in the process. Congress has frequently used its delegated powers to enact 'pre-emption statutes' that can eliminate or restrict any prior state regulatory policy in a particular area. In some instances pre-emption reflects a federal effort to eliminate state-by-state variation or set a federal standard well below that of the most aggressive states. Between 1995 and 2004 for example, a Republican Congress enacted 75 pre-emption statutes, many of which reflected 'pressure from business interest groups for the establishment of harmonious regulatory policies' across state boundaries (Zimmerman, 2005, 361). Pre-emption can also be used in instances where Congress deems states unable or unwilling to establish a needed regulatory presence, perhaps in instances in which state policy processes are deemed 'captured' by business and industry. In recent decades, pre-emption has been applied to a wide range of economic functions, from restricting state oversight of mutual funds to establishing uniform nutritional labelling on food items (Teske 2005). Any number of pre-emption options could hypothetically be applied in the climate arena and are most likely applicable in instances when the total number of active states are more limited or their policy implementation remains in early stages. This approach was clearly on the agenda of the 111th Congress.

Total pre-emption represents the most comprehensive form of federal encroachment, with the power to effectively eliminate all existing state laws and standards in a given area in favour of a uniform federal alternative. For example, Congress could simply declare through statute all 29 operational renewable portfolio standards null and void and install a national standard. It could take a similar approach in the cap-and-trade arena, shutting down state and regional carbon trading zones through a single legislative pronouncement. Some states and a number of state professional associations were clearly concerned that total (or near total) pre-emption could emerge as a federal strategy.

A somewhat early indicator of potential dividing lines over a total pre-emption strategy emerged in 2007 when senators from both parties introduced federal climate legislation that would eliminate all existing state policies that had been approved and prohibit future state efforts without explicit federal support. Some business and electric utility representatives strongly endorsed this approach, having pursued it in previous air quality regulation strategies. But intergovernmental organizations such as the National Association of Clean Air Agencies registered strong concerns and proposed other approaches. Senate sponsors ultimately backed down in the face of this strong resistance from state and local authorities, as well as environmental advocacy groups (NACAA 2008). The idea of total pre-emption of state climate policies re-emerged in the 111th Congress. However, no such policies were enacted and there appeared little chance of comparable legislation emerging from subsequent Congresses, reducing any prospects for federal domination via encroachment on earlier state initiatives.

Bargained Pre-emption: Negotiating a Federal-State Deal. There is also ample precedent for addressing federalism through extended bargaining between federal and state authorities. Rather than attempt an intergovernmental *coup d'état* through total pre-emption, the federal government can either negotiate terms directly with states or embrace their most ambitious efforts as a national standard. Such an approach may be particularly common when states are divided on an issue, allowing the federal government to broker differences and even offer side payments to potential losers. One early illustration of this dynamic was the 2009 Rose Garden ceremony that brought together the President, prominent governors and many other interests to launch a new agreement to reduce vehicular GHG emissions.

Federal law had long given federal authorities exclusive authority to set national fuel economy standards for vehicles and 2007 legislation included a modest increase in these standards. California, however, stepped aggressively into this very area through 2002 legislation (AB 1493) that exercised its unique powers to request a formal waiver from federal powers to approve proposed standards for carbon emissions from future vehicle fleets under the Clean Air Act. This reflected a long-standing intergovernmental agreement whereby California alone could exceed any federal policy on air emissions, as long as it secured federal approval. Only at that juncture could other states embrace California's approach as their own policy or simply accept whatever federal standards existed. This state effort involved a very creative reframing that essentially would have the effect of regulating fuel economy but doing so through the mechanism of air quality emission standards whereby California could retain jurisdiction with federal approval. This California initiative contributed to the decision by Massachusetts and other states to enter the courts seeking federal designation of carbon dioxide as an air pollutant, to force the reluctant federal hand on this issue.

The Bush Administration refused to budge on this, even after the Supreme Court decision, effectively blocking the California waiver request. But the incoming Obama Administration reversed course through an intergovernmental agreement that 'essentially plans to spread California's vehicle emissions restraints nationwide' through 2016 (Nivola 2010 162). This served to eliminate the possibility of two sets of standards (those endorsed by California and its state allies and those set federally and preferred by other states), thereby offering some predictability to vehicle manufacturers. California and like-minded states welcomed this outcome, as it in effect embraced their standards as national policy. This may explain why Schwarzenegger decided to spend a California election day with profound consequences for his state's fiscal future celebrating on the lawn of the White House. In turn, the Obama Administration built on this agreement to set even more ambitious targets through 2025 in direct collaboration with California officials.

Intergovernmental bargaining can also take other forms, blending regulatory mandates with financial incentives. During the 110th Congress, one prominent Senate cap-and-trade proposal, known as the Climate Security Act, featured a plan to offer financial incentives for 'clean state actions.' Under this proposal, those states that had 'led the nation' in attempting to reduce GHG emissions and improve energy efficiency would be eligible for substantial intergovernmental financial transfers through carbon emission allowances. In exchange the federal government would require that states discontinue involvement in any existing regional cap-and-trade programs and refrain from launching any additional programs on their own, thereby facilitating full state 'transition' into the new federal system. The collapse of Senate deliberation over climate legislation shortly thereafter prevented a serious hearing of this proposal but it demonstrated one way in which financial incentives might be used to attempt to broker an intergovernmental agreement.

A new version of this approach surfaced in the House in 2009, through the cap-and-trade provisions included in the American Clean Energy and Security Act (ACESA). This proposal would have pre-empted all existing state and regional cap-and-trade programs but only for the first five operational years of the federal program. Subfederal programs would presumably go into a deep freeze during this period, with the possibility that they might later be allowed to revive. In response, states would be granted direct oversight of some of the allowances that the federal government would allocate for the cap-and-trade program, though likely with less emphasis on rewarding early action states than under the Climate Security Act proposal. The overall allocation of allowances to the electricity sector was quite generous and state public utility commissions were given lead roles in determining distribution and oversight. All of this was part of a remarkably complex set of intergovernmental and inter-institutional negotiations designed to secure a majority vote in the House (Rabe 2010). This was achieved on a 219-to-212 vote on ACESA in June that generally followed partisan lines. Companion legislation never got far in the Senate, however, and there appeared to be virtually no prospect of revisiting this approach in the 112th Congress.

Partial Pre-emption: Setting a Federal Floor. Another approach that entails shared authority is partial pre-emption, which has been used extensively in water and air quality control as well as many non-environmental policy areas. Under this approach, there is less emphasis on using money as an intergovernmental lubricant and instead a willingness to sustain a two-tier system that allows for continued state policy implementation alongside a federal counterpart. Federal legislation may establish a form of 'floor pre-emption' that sets a national baseline that must be met in all states while allowing individual states to exceed that baseline by sustaining earlier programs or adopting new ones. Such an approach is particularly common when many states have adopted their own similar policies and are well into implementation, thereby making total pre-emption more contentious.

Many federal proposals over the last half-decade have focused on renewable electricity, including provisions to establish a national RPS. Most of these have followed a partial pre-emption strategy that would not interfere with existing state programs. Such legislation does not constrain existing or future state policies, including those with targets greater than the one proposed for the nation. There appeared to be some bipartisan support for returning to this model in the 2010s, particularly if the definition of 'renewable' would expand to include nuclear, natural gas and large hydro sources under the alternative frame of 'clean energy.'

Collaborative Federalism: An Intergovernmental Partnership. One additional option would view climate governance as an ongoing intergovernmental partnership rather than a formal distribution of power as would exist under pre-emption. There is some precedent for fashioning intergovernmental strategies expressly designed to engage both federal and state governments in a fairly collaborative manner that plays to their respective strengths. Under this model, the federal government often establishes an overarching framework that specifies required emissions levels, sets forth expectations of what states are to deliver in terms of results and maintains a mechanism to measure state performance with possible ramifications for failure. But the core of decision making on implementation in such instances is largely devolved to individual states. This would enable them to take advantage of the considerable expertise they have already developed in climate governance and tailor emission reduction strategies that are particularly well suited to their unique circumstances. A number of air quality programs operate under this type of framework, with states delegated considerable authority for decisions such as permitting and enforcement once EPA approves their 'emission budgets' and 'implementation plans.' There has also been considerable experimentation in this arena, particularly among states with a tradition for early engagement and capacity to work across traditional medium and functional boundaries. Given recent tensions between federal and state levels and the weakening of institutions designed to foster intergovernmental collaboration in the United States, such policies are not easy to launch or sustain.

Applied to climate change, a collaborative federalism approach might take two forms. First, the federal government could set emission caps for individual states or regions and then allow these sub-federal units to work out the necessary steps to implement existing policies or create new ones that would enable them to meet their reduction targets. States might receive federal financial incentives for meeting performance targets but lose access to these funds if they failed to deliver, perhaps building on the experience of the National Environmental Performance Partnership System launched in the 1990s. Second, programs such as federal cap-and-trade could allow states considerable latitude in implementation. This might follow the example of RGGI, whereby regional authorities determine the metrics for distributing emission allowances but defer to individual states to decide how to allocate them internally. This approach has some parallels to the European Union Emissions Trading Scheme. Under RGGI a multi-state organization headquartered in New York oversees the trading system and distributes allowances according to an established formula. But each state within RGGI then determines allocation (Cook 2010). These states have gravitated towards a system that auctions these allowances, generating considerable revenue and moving closer to what many economists have deemed the most effective way to operate cap-and-trade. In contrast, most serious Congressional proposals eschewed or limited auctioning in favour of a distribution process that would allocate most allowances free to industries in attempting to build a supportive coalition. One could indeed use the RGGI system as a national model on cap-and-trade, following the long history of federal borrowing of policy ideas from one or more states. Such ideas emerged in a number of policy briefs, including some produced by organizations representing state governments, but were generally marginalized in Congress.

However, one possible form of collaborative federalism may move forward in the absence of federal legislation, albeit amid much uncertainty and with considerable potential for conflict. The Obama EPA established an 'endangerment finding' for GHGs under 1990 air quality legislation, thereby launching a process to regulate emissions from large industrial sources and directly involve states through air permit implementation. This has proven highly controversial, with the possibility of federal legislation to deter this approach, litigation to block implementation or reversal under a future President. The EPA received a very mixed response from states as it moved into early stages of implementation. On the one hand, the lead environmental officials of 12 states petitioned Congress not to block EPA involvement and 16 states endorsed the agency's opposition to industry litigation to reverse the finding. On the other hand, 12 states, led by Alabama, Texas, and Virginia, petitioned EPA to terminate this regulatory process. The federal agency continued to move forward and ultimately seized control of the GHG permitting process in Texas when the state refused to cooperate.

The EPA repeatedly emphasized in 2010–2012 that it wanted to work in full partnership with states in developing this regulatory program. This included securing state input on program design as well as exploring the development of a permit fee structure that could transfer funds to states to cover implementation costs. The EPA has also worked with states to find ways to phase in GHG regulations over time, beginning with a set of very large emission sources and expanding that gradually in later years. The agency has faced considerable pressure from environmental groups to move more rapidly and cover more sources but senior officials have been adamant about the need for a 'manageable and commonsense' process. The longer-term development of this regulatory program remains highly uncertain and initial plans would still bypass many major emissions sources even at the end of the phase-in process. But this represents, at least initially, an effort between federal and state regulatory agencies to jointly craft an implementation plan and also avoid interference with a wide range of existing state policies. Some officials from states active in climate policy requested the EPA to allow them to develop locally and regionally tailored compliance plans that would include use of existing state programs to meet new federal GHG emission reduction requirements.

Back to the States? 2012 and Beyond. The limited scope and uncertain future of new federal climate policy initiatives under contested federalism underscored the reality that much of American climate governance will likely retain a strong state and regional flavour in the coming years. After the surge of sub-federal policy development in the period of state domination, states began to slow their efforts, in large part due to anticipated federal action on a large scale. The collapse of Congressional deliberation on major legislation returned much of the lead in climate governance to states. This raised significant questions of implementation, including a series of major challenges and opportunities.

States shifted their policies into full operation at a point of enormous fiscal constraint, leaving uncertain their capacity to sustain the levels of staff support necessary to assure effective implementation. In turn, their precarious economies may have made them more reluctant to move fully into implementation given the possibility of imposing added costs for energy and manufacturing. Some have begun to question their capacity and willingness to sustain earlier commitments, particularly in cases of changes in political leadership after 2010 elections. In the Western Climate Initiative, all states except California eventually withdrew by 2012. Even California, the driving force behind the WCI process, faced a November 2010 ballot proposition that, if enacted, would have placed its own commitment to cap-and-trade in jeopardy. That initiative was rejected by a 61 to 39 per cent margin but California faced numerous challenges in moving ahead, including litigation over possible uses of the auction revenues. These kinds of shifts may be further propelled by major swings in public opinion on the existence of climate change and the anticipated severity of any potential risks. Public understanding of this large-scale phenomenon appears highly sensitive to short term considerations, such as storm intensity, drought, and temperatures in a given summer or winter. This appears to produce rapid short-term fluctuation in attitudes to the existence of climate change as well as support for various state policies (Borick and Rabe 2010).

At the same time, three new developments emerged in recent years that could serve to bolster existing state commitments or even expand them. First, the RGGI experience with auctioning allowances produced a significant source of revenue in concert with imposing a carbon cap. More than $1 billion was returned to the ten RGGI states via their first four years of auctions, with most of this funding channelled directly to renewable energy and energy efficiency initiatives. The RGGI model received significant attention in other states given its ability to impose a carbon price and thereby diversify a state's fiscal base. Other states increasingly explored a range of tax and fee options that might borrow from this experience.

Second, the emergence of Canadian provinces into a more active mode of climate policy development has expanded the opportunity for cross-border partnerships with American states. Provinces became active partners in the regional cap-and-trade programs, although not all remained active into implementation. Some have also developed a pair of policies that have been frequently discussed in the United States but never enacted in any state. Carbon taxes that establish a levy based on the carbon content of various fossil fuels have been enacted in British Columbia and Quebec; 'feed-in tariffs' that guarantee long term prices above what is customary for conventional energy sources have been established in Ontario (Burke and Ferguson, 2010).

Third, a growing number of states have begun to consider policies designed to facilitate adaptation to changing climate, in some instances alongside efforts to attempt to mitigate climate change through emissions reduction. Twelve states that border an ocean had completed state adaptation planning processes by 2012, reflecting significant concern about rising ocean levels and potential impact on coastal land use. Water scarcity through inland drought has also emerged as a growing concern, particularly in the Southwest. Thus the issue of adaptation is clearly gaining increasing attention, particularly as states and regions respond to climatic changes that are consequential in their areas. This could lead to new policy departures in such areas as insurance and land use planning.

CONTESTED FEDERALISM AND BEYOND

The transition from state domination to contested federalism in American climate change policy remains at a fairly early and uncertain stage. The extended period of state experimentation with the development and implementation of numerous climate policy tools creates a remarkable testing ground to examine what does – and does not – work effectively. This affords a unique opportunity for intergovernmental 'policy learning,' discerning lessons from best practices that might be blended into a system that engages both states and federal units in a constructive manner. Such a process would be entirely consistent with Douglass North's understanding of 'adaptive efficiency,' which reflects 'the willingness of a society to acquire knowledge and learning, to induce innovation, to undertake risk and creative activity of all sorts, as well as to resolve problems and bottlenecks of the society through time' (North 1990, cited in Bednar 2008). One could envision an iterative process that builds on respective federal and state capacities to generate a viable governance regime to address the daunting challenges of climate change (Carlson 2009).

Thus far however, there has been little evidence that this will occur. Much of the focus on climate change in the 111th Congress involved a desperate search for sufficient votes to assemble a lengthy and complex bill. States were one of many claimants for most-favoured treatment under such a massive bargaining umbrella. This process stalled leaving a patchwork of federal and state policies. More than two decades after the enactment of the 1990 air quality legislation, Congress has issued no formal statement on whether it deems GHGs an air pollutant or if anything should be done to reduce them. Federal government engagement on this issue has been dominated by the judicial and executive branches and is susceptible to much future modification. One major test of federal ability to forge a constructive partnership with states in a period of contested federalism will involve EPA's attempts to implement its climate endangerment finding through GHG permits that will ultimately be issued by state agencies. This moves forward, however, amid potential challenges from the courts, Congress, states such as Texas, and presidents who follow Barack Obama. If this federal effort is blocked and others do not follow, it is possible that we may shift back from contested federalism to state domination of American climate policy in coming years.

States have increasingly moved beyond experimentation into the stage of implementing their own climate policies. What once seemed a short-term experiment that would eventually be eclipsed by far-reaching federal legislation continues to occupy centre stage in American climate policy. Such core climate policy tools as cap-and-trade and RPS continue to operate only at the sub-federal level, albeit amid shifts in political leadership and in a very uncertain fiscal period for states to sustain existing efforts. In turn, the experience with cap-and-trade in the North-east and California raises the question of whether states may expand their involvement in carbon pricing strategies, thereby linking efforts to address climate change with a mechanism to diversify their revenue bases. In any event, American climate policy will likely retain a significant state imprint in the coming years, barring some major state withdrawal from this arena. This remains ironic given the earlier framing of climate change as a matter exclusive to national governments and international regimes. This pattern is also evident in many other nations around the world that allow for intergovernmental sharing of authority for energy, environmental protection and all other policy areas related to climate change.

REFERENCES

Bednar, Jenna (2008), *The Robust Federation*, Cambridge: Cambridge University Press.

Bomberg, Elizabeth (2009), 'Governance for Sustainable Development,' in Miranda Schreurs, Henrik Selin, and Stacy VanDeveer (eds), *Transatlantic Environment and Energy Policies*, London: Ashgate, pp. 21–40.

Borick, Christoper, and Barry Rabe (2010), 'A Reason to Believe: Examining the Factors that Determine Individual Views on Global Warming,' *Social Science Quarterly* **91** (September), 777–800.

Burke, Brendan, and Margaret Ferguson (2010), 'Going Alone or Moving Together: Canadian and American Middle Tier Strategies on Climate Change,' *Publius* **40** (Summer), 436–59.

Carlson, Ann (2009), 'Iterative Federalism and Climate Change,' *Northwestern University Law Review* **103**, 1097.

Cook, Brian (2010), 'Arenas of Power in Climate Change Policymaking,' *Policy Studies Journal* **38**, 465–86.

Engel, Kirsten (2009), 'Whither Subnational Climate Change Initiatives in the Wake of Federal Climate Legislation?' *Publius* **39** (Summer), 432–54.

Farber, Daniel (2008), 'Climate Change, Federalism, and the Constitution,' *Arizona Law Review* **40**, 879.

Hoffmann, Matthew (2011), *Climate Governance at the Crossroads,* New York: Oxford University Press.

Kysar, Douglas, and Bernadette Meyler (2008), 'Like a Nation State,' *UCLA Law Review* **55**, 1621.

Lowry, William (1997), *The Dimensions of Federalism*, revised edition, Durham: Duke University Press.

Lowry, William (2008), 'Disentangling Energy Policy from Environmental Policy,' *Social Science Quarterly* **89** (December), 1195–211.

NACAA (2008), 'Defining the Role of States and Localities in Federal Global Warming Legislation', *Discussion Papers* (Washington, DC: NACAA).

Nathan, Richard (2008), 'Updating Theories of Federalism,' in Timothy Conlan and Paul Posner (eds), *Intergovernmental Management in the 21st Century*, Washington, DC: Brookings Institution, pp. 13–25.

Nichols, Mary (2013), 'De-Carbonizing California is the Path to Progress and Innovation', GAME CHANGERS 2013 (San Jose, CA: Silicon Valley Leadership Group, 2013), p. 38.

Nivola, Pietro (2010), 'The Long and Winding Road: Automobile Fuel Economy and American Politics,' in Barry Rabe (ed.), *Greenhouse Governance*, Washington, DC: Brookings Institution, pp. 158–80.

North, Douglass (1990), *Institutions, Institutional Change, and Economic Performance*, Cambridge: Cambridge University Press.

Pendergrass, John. 2010. "Texas Tailors the Tailoring Rule," *Environmental Forum* (September-October): 10.

Peterson, Paul (1995), *The Price of Federalism.* Washington, DC: Brookings Institution.

Pooley, Eric (2010), *The Climate War,* New York: Hyperion.

Posner, Paul (2010), 'The Politics of Vertical Diffusion: The States and Climate Change,' *Greenhouse Governance*, pp. 73–100.

Rabe, Barry (2004), *Statehouse and Greenhouse: The Emerging Politics of American Climate Policy*, Washington, DC: Brookings.

Rabe, Barry (2008), 'States on Steroids: The Intergovernmental Odyssey of American Climate Change Policy,' *Review of Policy Research* **25** (2), 105–28.

Rabe, Barry (2010), 'The Aversion to Direct Cost-Imposition: Selecting Climate Policy Tools in the United States,' *Governance* **23** (4), 583–608.

Selin, Henrik, and Stacy VanDeveer (eds) (2009), *Changing Climates in North American Politics,* Cambridge: MIT Press.

Teodoro, Manuel (2011), *Bureaucratic Ambition: Careers, Incentives, and the Innovative Administrator*, Baltimore: Johns Hopkins University Press.

Teske, Paul (2005), 'Checks, Balances, and Thresholds: State Regulatory Re-enforcement and Federal Pre-emption,' *PS: Political Science and Politics* **38** (July), 367–70.

Volden, Craig, Michael Ting, and Daniel Carpenter (2008), 'A Formal Model of Learning and Diffusion,' *American Political Science Review* **102** (August), 319–32.

Zimmerman, Joseph (2005), 'The Nature and Political Significance of Preemption,' *PS: Political Science and Politics*, **38** (July), 359–62.

8. Institutional strength, intergovernmental relations, and national climate policy coordination: Australia and Canada compared

David Gordon and Douglas Macdonald

The nature of climate change is such that no one state can solve the problem by itself, nor is any one state likely to incur the costs of emission mitigation without some assurance of comparable action by others (Olson 1965; Hardin 1968). This same dynamic is replicated in federal states: How can one attain joint objectives and coordinated actions from autonomous yet interdependent jurisdictions in the face of considerable incentives to defect? As of 2013, Canada has been unable to reach agreement on national objectives, policy mechanisms, or the allocation of costs between sources and Canadian provinces and has relied instead on uncoordinated, unilateral actions by the federal government and provinces. By contrast, in 2011 Australian jurisdictions agreed upon a coordinated plan, including national strategies and objectives on renewable energy, energy efficiency, adaptation and mitigation. How do we explain Australian success in achieving coordination, in contrast to Canadian failure over the same time period? While differences in political culture, geography, political economy and political leadership no doubt play a part, we suggest that the institutions of intergovernmental relations (IGR) in each country have had a significant effect on efforts to coordinate.

Our objective is to explore whether there are meaningful differences in the institutions of IGR that create permissive or constraining conditions in these two countries. The chapter begins by situating our analysis within the literature on federalism, intergovernmental relations and multilevel governance. We then set out our comparative framework and describe national coordination efforts, focusing on the institutions of IGR and on coordination outcomes that have been realized. The subsequent section applies the conceptual framework and offers some thoughts regarding the mechanisms through which these differences might be linked to coordination outcomes.

FEDERALISM, MLG AND INSTITUTIONAL RESPONSES TO INTERDEPENDENCE

Federal states in Elazar's definition (1987) embody some combination of 'self plus shared rule.' Applied to modern governance and public policy, this translates into increasingly complex dynamics of interdependence and autonomy, giving rise to an overlap between studies of federalism and multilevel governance (Cameron and Simeon 2002; Painter 2001). This functional interdependence means that each level of government in a federal state has become reliant on others to achieve its policy objectives (Braun 2006), a dynamic that is especially apparent in the issue area of climate change. As with the international level, in federal states emissions reductions made in one jurisdiction are meaningless if they are offset by increased emissions in another. Furthermore, there are powerful incentives for jurisdictions to avoid costly efforts to reduce emissions since those costs are direct and discrete (in economic and electoral terms), while the benefits they produce are diffuse and difficult to quantify (Ostrom 1990; Harris 2007).

The disincentives to cutting emissions are compounded by the division of powers in federal states such as Canada and Australia. The linkage between emissions and economic activity (agriculture, forestry, mining, oil and gas recovery and refining, manufacturing, power generation) creates tension between, and within, federal and sub-federal jurisdictions. In Canada and Australia, the provinces/states (not the national government) are the owners of natural resources within their borders. This creates strong incentives for resource extraction and exploitation, as well as resistance to central interference and interdependence in terms of achieving common environmental objectives. How then to coordinate effectively in the face of such barriers?

One possible approach is to centralize and impose hierarchical governance by an aggressive assertion of federal authority (Weibust 2009; Hardin 1968). However, the condition of functional interdependence leads to serious questions regarding the federal ability to effectively govern through centralization (Keating and Wanna 2000; Cameron and Simeon 2002) and the declining legitimacy of shared-cost program funding of the sort that characterized Canadian federalism in the 1950s and 1960s suggests an eroded ability of the federal government to unilaterally dictate the process (Meekison et al. 2004). Resistance by subnational jurisdictions to federal intrusion, due to powerful structural and electoral incentives, is a further and substantial barrier to centralization (Bolleyer 2009). Policy coordination can also take place outside institutional channels, either through arms length emulation or bilateral interaction (Bolleyer and Börzel 2010). However such approaches are beset by substantial transaction costs and run the risk of producing incoherent objectives and policies.

This opens the door to collaboration as a process for coordinating federal and subnational governments to produce coherent national objectives, strategy and instruments for climate change mitigation. While such collaboration does bring with it the risk of lowest common denominator outcomes (Scharpf 1997), recent work suggests the potential importance of 'non-hierarchical coordination' (Bolleyer and Börzel 2010), constrained collaboration (Braun 2006) and complex multilevel governance (Selin and VanDeveer 2009). While differing along some dimensions, all acknowledge that effective collaboration requires institutional venues in which jurisdictions can interact and engage in efforts at intergovernmental problem solving.

We define effective coordination as the mutual agreement of jurisdictions on joint objectives *and* joint efforts to achieve them – both what to do and who does what. Our definition focuses on policy outputs, rather than directly on outcomes of those policies (e.g. actual emissions reductions or increases in renewable energy production/consumption). This distinction between outputs and outcomes assumes governments are unable to exert direct control over emissions (regulations and price mechanisms can be used, but emissions are ultimately produced through the individual and aggregate choices of individuals and organizations) and, recognizing the multiple intervening factors that operate between output and outcome, focuses attention on the former (Montpetit 2003).

COMPARATIVE FRAMEWORK

Applying institutional analysis to the challenge of federal policy coordination, we pose two questions. (1) Does variation in the institutions of IGR explain why Australia was able to coordinate whereas Canada was not? (2) If yes, what are the mechanisms that link institutional strength to coordination? In order to address these, we will attempt to measure and compare institutional strength in our two cases, and advance some conjectures regarding the underlying causal mechanisms at work.

We define institutions of IGR as the rules of the game that structure, enable and constrain interactions amongst jurisdictional actors (North 1990). They may be formal or informal and relate to participation, agenda setting, decision making, distribution and enforcement (Painter 1991). In addition, institutions often include the creation of a supportive bureaucracy, which may facilitate coordination. For example, intercantonal institutions for Swiss energy policy 'do not have means of coercion, but they create an identity and produce norms' which impact policy diffusion (Strebel 2011). Similarly both Bolleyer (2009) and Weibust (2009) assess the impact of rules *and* institutional bureaucracy on efforts at intergovernmental coordination. Secretariats, even without rule-making authority, can influence governments, according to the authors cited above.

We define institutional 'strength' as the ability to influence the actions of participating governments and we examine both the rules and the institutional bureaucracy as sources of strength or weakness. We suggest that the ability of rules to influence governments toward collaboration has three sources: (1) the characteristics of the rule itself; (2) the degree of codification and publicity; and (3) enforceability of the rule. Rules are expected to increase institutional strength if they provide for regularity and predictability, entrench majority or qualified majority voting rather than consensus (Bolleyer, 2009; Brown 2003) and are broadly applicable (to both federal and subnational governments). Furthermore, rules are expected to increase institutional strength as they are increasingly codified and specified (Bolleyer, 2009) and publicized and archived (Poirier, 2004). Lastly, rules are expected to increase institutional strength as a function of effective enforcement mechanisms (Simeon and Nugent, 2008).

As a source of institutional strength, bureaucracy can be assessed in terms of two factors: (1) mandate and (2) resources (Biermann, 2001). We expect institutional strength to increase with greater bureaucratic autonomy and clarity of mandate (Bolleyer and Bytzek 2009). Similarly we expect institutional strength to increase as a function of resource capacity, including financial resources (amount, sources, predictability) and capacity (staffing). Table 8.1 provides a summary presentation of variables expected to impact institutional strength.

CASE SELECTION

We focus our analysis on Canada and Australia since, as of 2012, Australia had successfully coordinated national climate policy whereas Canada had not. These countries share many economic, geographic and political similarities. Both are rich in natural resources and depend on extractive industries and export markets. Both combine expansive

Table 8.1 Elements contributing to institutional strength/eeakness of IGR

	Weaker	Stronger
Rules	No rules	Rules exist (informal or formal)
	Ad hoc meetings	Regular meetings
		Predictable/scheduled meetings
	Consensus decision-making	Majority or QMV voting
	Limited participation/ applicability	Broad participation/ applicability
	No constitutional basis for IGR	Constitutional basis for IGR
	No legal basis for agreements	Agreements embedded in legislation
	No public release of agreements	Public statements released regarding agreements
	No third party enforcement exists	Third party enforcement or oversight exists
Bureaucracy	No official mandate	Official mandate
	Limited or unpredictable financial resources	Greater and regularized financial resources
	Limited staffing or reliance on short term secondments	Dedicated and long-term staffing

Source: the authors

territory with dispersed yet urbanized settlement. In terms of climate change both Canada and Australia are among the world's highest per capita emitters; each accounts for approximately 2 per cent of global greenhouse gas (GHG) emissions.

Each has a federal constitution setting forth the distribution of powers between levels of government. Regarding climate change, the constitutional allocation of powers and concurrent jurisdiction in both Canada and Australia result in functional interdependence over environmental policy because no level of government has exclusive authority (Keating and Wanna 2000; Brown 2003). Each has a relatively weak national legislative institution of regional accommodation (Senate) and relies extensively on IGR outside elected bodies to manage interdependence between the federal and subnational levels of government.

There are also important differences. The deep cultural-historical-regional cleavages that exist between Quebec and the rest of Canada (and to some extent, between Western and Eastern Canada) do not exist in Australia. While Canadians hold a range of often incompatible views on Canadian federalism (e.g. Canada as an equal partnership of provinces, a single entity or an amalgam of two distinct nations), the federal concept is comparatively unproblematic in Australian politics (Braun 2006; Brown 2003). Australia is the more centralized federation of the two (Thorlakson 2003) as a result of higher vertical fiscal imbalance and the tendency of the High Court to rule in favour of the Commonwealth in matters related to jurisdictional conflict (Painter 1998; Braun 2006).

Political parties are more tightly integrated between federal and subnational levels in Australia than Canada, which may contribute to greater coordinating capacity in times of party overlap (Bolleyer and Bytzek 2009; cf. Painter 2001). There is evidence that party dynamics have contributed to vertical and horizontal coordination in Australia. Periods such as the creation of COAG in the early 1990s and the 2007–2009 episode of national climate policy coordination occurred when the Labor Party controlled both the Commonwealth government as well as a substantial number of state governments. In Canada, although federal and provincial parties often have the same name, each provincial party is largely independent from its provincial and federal namesakes.

There are also significant differences between the two nations in terms of climate change policies. Under the Kyoto protocol, Australia's target (an 8 per cent increase over 1990 levels by 2012) was far more lenient than Canada's (a 6 per cent reduction from 1990 levels by 2012). Thus Canada faced a 31 per cent emissions reduction from a Business as Usual scenario (BAU), compared to a 19 per cent cut for Australia (Harrison and Sundstrom 2007). In parallel with Canada's regional economic cleavages, compliance costs between Australian states and territories vary much less than between Canadian provinces, where Alberta's projected costs could be ten times as high as those of other provinces (see Table 8.2). Australia has more directly experienced climate change than has Canada, with both symbolic effects (degradation of the Great Barrier Reef) and extended drought and extreme flooding in high population and economically important regions (Crowley 2007). In Canada, the greatest changes have been in the North, of which the rest of Canada has little awareness. Lastly, Australia has a much more diffuse set of trading relationships than does Canada, where reliance on a single export market (the United States) has led to substantial pressure to conform to United States (US) federal climate policy (Brown 2003).[1]

Table 8.2 Projected impact on jurisdictional GDP relative to business as usual (BAU) scenario in 2020

GDP change relative to BAU (2020), ranked from worst to best			
Canada (20% emissions reduction from 2006 levels)		Australia (5% emissions reduction from 2000 levels by 2020)	
Alberta	–7.3%	Queensland	–1.7%
British Columbia	–2.2%	New South Wales	–1.5%
Saskatchewan	–1.2%	Victoria	–1.4%
Quebec	–0.7%	South Australia	–0.5%
Atlantic provinces (Nova Scotia, New Brunswick, Prince Edward Island, Newfoundland and Labrador)	–0.5%*	Tasmania	–0.5%
Ontario	+0.6%	Western Australia	+0.1%
Manitoba	+1.9%	Northern Territory	n/a
Northwest Territories, Yukon, Nunavut	n/a		

* *The projections for the Atlantic provinces are grouped together rather than modelled individually.*

Sources: Bramley et al. 2009: 12. The calculations for Canada are based on a 20% reduction by 2020 from 2006 levels. Access Economics 2009a: 15-16. The calculations for Australia are based on the CPRS target of a 5% reduction by 2020 from 2000 levels. Similar modelling was conducted by the Commonwealth Department of Treasury, with results much the same. See Australia 2008: 161

Recognizing that the aforementioned factors all affect attempts at coordination in our two cases, we nonetheless argue that variations in institutional strength are part of the explanation. In 2011 Australia successfully adopted a national policy, after many years of fruitless negotiation while, as of 2012, Canada had no coordinated federal-provincial mitigation strategy. The causal factors cited above changed relatively little before and after 2011, suggesting that institutions may play a critical role in explaining Australia's eventual success. The following two sections summarize efforts at coordinating national policy in each country, highlighting the institutions of IGR used and the outcomes achieved.

AUSTRALIAN IGR AND NATIONAL CLIMATE POLICY COORDINATION

Two institutions of IGR have served as fora for collaboration towards a national climate change policy in Australia. First, the Council of Australian Governments (COAG), created in 1992, is an extra-constitutional institution of IGR intended to facilitate coordination between members of the federation and to oversee the existing set of Joint Ministerial Councils (Painter 1998; Carrol and Head 2009). COAG is organized in tiers, with First Ministers from the Commonwealth and all state/territorial governments comprising the top layer, a COAG Senior Official committee (senior bureaucrats) comprising the second layer and a set of subsidiary committees forming a third layer (Carrol and Head 2009). Council of Australian Governments meetings are called and chaired by the Prime Minister, who also sets the agenda. The COAG permanent secretariat is housed within the Commonwealth Department of the Prime Minister and Cabinet. All decisions made within COAG are on the basis of unanimity.

The second major institution of IGR is the Council of the Australian Federation (CAF) which consists solely of state and territorial governments, excluding the Commonwealth government. The Council of the Australian Federation has its origins in early efforts at horizontal coordination among the states and territories in order to limit Commonwealth dominance during initial COAG negotiations. In October 2006, CAF was created by an intergovernmental agreement (IGA) (signed by all states and territories) to institutionalize horizontal coordination and to produce coherent, joint state-territorial positions prior to COAG meetings (CAF 2006; Tiernan 2008). The Council of the Australian Federation was created as a complement to COAG, but was also envisioned as an alternative forum for national coordination in the absence of Commonwealth engagement or leadership (ibid).

The Council of the Australian Federation is chaired on a rotating basis by the state/territorial First Ministers and is mandated to meet at least once a year, although such meetings are often supplemented by out of session interactions (ibid). The Council of the Australian Federation is supported by a permanent secretariat that is jointly funded by the member states and territories and is located in the Chair's jurisdiction. The rotating Chair sets the agenda, in consultation with other First

Ministers, but decisions are made unanimously (CAF 2006). Meetings are complemented by forward work plans and required progress reports on actions agreed upon in prior meetings (Tiernan 2008).

EVOLUTION OF NATIONAL CLIMATE POLICY

Early efforts to develop a national approach to climate change took place from 1990–1994. Christoff (2005) characterizes this as a period of 'naive altruism,' rooted in ideal driven agreements on national targets and measures. In 1992 the *National Greenhouse Response Strategy*, consisting of voluntary measures, was released. It was officially endorsed by COAG, indicating some involvement by First Ministers.

In 1996 the election of John Howard's Liberal Commonwealth government led to a renewed emphasis on voluntary measures and a shift away from specific targets (Crowley 2007). The Howard government sought to protect Australian economic competitiveness and avoid the imposition of 'unfair' costs. The government went to the Kyoto conference with a hard line position on targets and demands that one time land use change emissions be included in their baseline. Both demands were met and Australia signed the Kyoto Protocol, accepting a target of an 8 per cent increase (Lyster 2004; Christoff 2005).

There is little evidence of any ongoing process of IGR during this period. From 1996 to 2001, COAG was convened only four times and the communiqués make little mention of climate change (Carrol and Head 2009). The Howard government had little interest in continuing earlier efforts at collaboration with the states and territories. It put IGR on a more bilateral, competitive and antagonistic footing (ibid). The Council of Australian Governments endorsement of the Commonwealth negotiating position for Kyoto indicates that the states/territories likely agreed on overall terms, but it is unclear whether there was a jointly agreed upon position regarding acceptable targets (COAG 2007b).

States and territories began experimenting with climate policy in the late 1990s. The Commonwealth removed itself from the international process by signalling its intent not to ratify the Kyoto Protocol in 2002. This led to a new dynamic of horizontal IGR. In 2004 the states and territories agreed to explore a national carbon emissions trading scheme (ETS) developed in the absence of federal leadership (NETT 2006). A joint taskforce of state and territory officials, the National Emissions Trading Taskforce (NETT), was established to investigate the design of a cap and trade system involving all states and territories (Ayers 2006).

This set the stage for CAF as an institution for coordinating state/territory positions on climate change (Tiernan 2008). From its inception, CAF adopted a strong position, issuing a joint communiqué in 2007 on the need to address climate change and the intention of the states and territories to develop a national ETS. The Council of the Australian Federation asserted a coherent and unified voice on behalf of the states and territories, strongly supporting the need for coordinated national action on climate change and it declared emissions trading to be the most appropriate means through which to reduce emissions. It also adopted a leadership stance and stated that it would proceed with a national emissions trading scheme should the Commonwealth itself fail to do so (CAF 2007b).

In response, Prime Minister Howard created, in 2005, a group of federal and business actors (the Task Group on Emissions Trading) and also told a senior officials' working group in COAG to explore possibilities for coordination between the states/territories and Commonwealth (Ayers 2006; COAG 2005). The COAG exercise may not have represented true engagement with the states/territories, however, as climate change was at the bottom of the agenda and received only cursory discussion (Hollander and Patapan 2007).

Despite this, vertical coordination began to emerge. In early 2006 COAG announced agreement on a new *National Climate Change Plan of Action*, a series of principles regarding national climate policy that carefully sidestepped the ongoing tensions regarding a national emissions trading system (COAG 2006). The senior officials' working group began to develop a national adaptation strategy and revisions to the national renewable energy framework. Later in 2006, COAG agreed on the need for a new national GHG reporting system, leading to agreement on the *National Greenhouse and Energy Reporting Act 2007*.

In November 2007 Labor Party leader Kevin Rudd was elected Prime Minister. He immediately reengaged with the states and territories through COAG, committing to meeting four times in 2008. A joint working group on climate change and water (WGCCW), drawn from the Ministerial Councils on energy and environment, was immediately formed under COAG and was tasked with incorporating the National Emissions Trading Scheme (NETS) and the recommendations of the Task Group on Emissions Trading into a national plan (COAG 2007a). In addition sub-groups within the WGCCW on renewable energy, energy efficiency and adaptation were created and tasked with designing new national policies in each area.

The Commonwealth government released a Green Paper on the Carbon Pollution Reduction Scheme (CPRS) in July 2008, and in October of the same year COAG announced agreement on principles regarding the fit between state/territory and Commonwealth mitigation measures and approaches to rationalizing existing policies and programs with the new CPRS (Wilkins 2008). It also announced agreement on a national approach to carbon capture and storage and the joint creation of the Global Carbon Capture and Storage Institute (COAG 2008b). In April 2009 COAG agreed on an updated national renewable energy target of 20 per cent by 2020 and later in the year on a new national energy efficiency strategy (COAG 2009b). In July 2009 COAG agreed to revise the Australian Energy Market Agreement in order to allow pass-through of costs associated with CPRS and Renewable Energy Target (RET) to regulated users (COAG 2009a).

From late 2007 to late 2009, COAG seems to have taken for granted the appropriateness of the CPRS as the national approach to reducing emissions. The WGCCW was explicitly given the job of reaching agreement on a national ETS, in addition to developing national policies on renewable energy, energy efficiency and adaptation. In response to vocal resistance over cost imposition, which states and territories (along with various industry organizations) perceived to be unfair, near term targets were relaxed and the permitting system under CPRS was made more flexible. The Commonwealth earmarked substantial subsidies to impacted sectors and regions (Taylor 2008; Berkovic and Hart 2008). In spite of this wrangling over the details of the trading scheme, the states, territories and Commonwealth agreed in March 2008 on the need for a national ETS, and communiqués issued in October 2008, as well as April and July 2009 (following national consultation on the CPRS green and white papers), present implementation of CPRS as taken for granted (COAG 2008b, 2009a, c).

However, CPRS implementing legislation was ultimately defeated (twice in 2009) in the Australian Senate, as a result of rejection by the opposition Liberal Party and the unwillingness of independent and Green Senators to support the government position. In spring of 2010, Julia Gillard replaced Rudd as Labor Party Leader and Prime Minister. The Carbon Pollution Reduction Scheme was subsequently reconfigured into a hybrid carbon tax/trading system (Commonwealth of Australia 2011). Support for this scheme amongst Australian states and territories has waned and there is a sense that the national policy coordination process has moved away from COAG and has been further centralized (Salusinszky and Barrett 2011; Salusinszky 2011). Several states, notably New South Wales, Western Australia, Queensland and Victoria, have publicly expressed displeasure at the imposition of a national carbon tax and have highlighted the declining role of COAG in coordinating the policy response (Kenny 2011). These actors are calling for compensation from the Commonwealth (Vasek 2011; McKenna and Salusinszky 2011; Packham 2011). There has been public bickering over the projected economic costs of the carbon tax (Wade 2011; Hurst 2011). In spite of state resistance, the carbon tax package was approved by both Houses of Parliament in the fall of 2011 and was set for implementation in June 2012 (Johnston 2011).

Australia thus offers a case of successful, though limited, national climate policy coordination with national targets for cutting GHGs and for increasing renewable energy. These targets were endorsed by both levels of government and were supported by jointly endorsed strategic documents (on energy efficiency, adaptation, carbon capture and storage), modified national legislation (renewable energy) and funding commitments (the climate action fund). In addition there was joint agreement within COAG on the CPRS package of legislation, indicating a successful process of allocating costs and efforts amongst jurisdictions. Since 2011 several developments, including the Commonwealth decision to reformulate national mitigation policy in the form of a hybrid carbon tax/emissions trading scheme, have revealed schisms between states and the Commonwealth that may undermine national policy coordination. However the discourse has been focused primarily around compensation and process; there remains broad commitment amongst the states and territories regarding the national target itself.

CANADIAN IGR AND NATIONAL CLIMATE POLICY COORDINATION

Executive federalism – policy making by negotiations between the federal Prime Minister and provincial Premiers – has been a key form of governance in many areas of Canadian policy. This form of IGR is practised through several institutions, the most important being the First Ministers Meeting (FMM). Sometimes referred to as the '"pinnacle" of Canadian IGR,' the FMM is an ad hoc institution convened by the Prime Minister, who sets the agenda (Papillon and Simeon 2004). The FMM has no codified or well-accepted rules regarding participation and attendance (ibid). Decision rules are also uncodified, although the unanimity rule is assumed. Because it has no permanent secretariat, FMMs cannot be considered to have autonomous bureaucratic support (ibid). Most importantly no FMM has ever been convened to explicitly consider Canada's policy response to climate change.

In contrast to the FMM, two other institutions of IGR did serve as fora for federal-provincial collaboration on national climate change policy between 1993 and 2002. The first of these is the Joint Ministers Meetings (JMM), established by two longstanding federal-provincial secretariats (the Canadian Council of Ministers of the Environment (CCME) and the Council of Energy Ministers (CEM)) after an initial federal-provincial plan of 1990 and Canada's ratification of the United Nations Framework Convention for Climate Change in 1992. The JMM was created to develop federal-provincial consensus on climate change and coordinate policy on emissions mitigation, bringing together all Ministers of Energy and of Environment on an annual or semi-annual basis. The JMM operated on the (uncodified) basis of consensus decision making and was supported by working groups comprised of senior level officials, including the National Air Issues Steering Committee (NAISC) and the National Air Issues Coordinating Committee (NAICC), both established in 1993. They had the mandate to address all air pollution issues, including climate change. Subsequently, the JMM designated a body of federal and provincial bureaucrats focused on climate change, the National Air Issues Coordinating Committee – Climate Change (NAICC-CC). The NAICC-CC was co-chaired by Alberta and the federal government and it in turn established seven issue specific working groups.

The second major institution of IGR was the National Climate Change Process (NCCP), coordinated by the National Climate Change Secretariat (NCCS), which was co-chaired by Alberta and Canada. The NCCS was created in April 1998 by the JMM in order to facilitate federal-provincial coordination, and was housed within the federal government's Climate Secretariat (itself created in February 1998 to help coordinate policy across federal departments). The National Secretariat had a dedicated (albeit seconded) staff of federal and provincial officials and the two secretariats shared an annual budget of 50 million CAD (Oulton 2010).[2]

The NCCP, which ran from 1998 to 2002, continued the institutional configuration of the JMM described above. The decision rule (again uncodified) in the NCCP was consensus. From 1998 to 2002 a series of 16 issue tables, each comprised of both government and non-government experts, developed policy options.

The Council of the Federation (COF), an exclusively provincial-territorial body of horizontal IGT, was created in 2003. In structure, it is equivalent to Australia's Council of the Federation. However, it has contributed far less to national climate policy making than its Australian counterpart.

HISTORY OF NATIONAL CLIMATE COORDINATION

In 1990 the federal government, under Progressive Conservative Prime Minister Brian Mulroney, unilaterally committed to stabilizing GHG emissions. The Mulroney government actively supported the United Nations Framework Convention on Climate Change, ratifying it in 1992. Following these commitments the federal government in 1993 established the JMM and began to work with the provinces to develop a national climate program. The first result was the 1995 National Action Program on Climate Change, a federal-provincial program. Its principal component was a joint government-business program intended to encourage voluntary emission reductions, the Voluntary Challenge and Registry (VCR).

Ministers participating in a JMM held in Regina on 12 November 1997 reached agreement that Canada's negotiating position at Kyoto should be stabilization of emissions at 1990 levels by 2012 (Stilborn 2003) (Quebec abstained, holding out for a more stringent target). Subsequent to the Regina JMM, the federal government, now under Liberal Prime Minister Jean Chrétien, unilaterally changed the national position to a *reduction* below 1990 levels, eventually agreeing to a 6 per cent reduction target. By ignoring the JMM decision and substituting its own target, the federal government caused considerable provincial mistrust (Harrison 2007; confidential interviews).

Following the Kyoto negotiations, a First Ministers Meeting, convened to address other issues, provided an opportunity for First Minister discussion of national climate change policy (Papillon and Simeon 2004; Stilborn 2003). The tension between the federal government and provinces was resolved by agreeing to reengage the federal-provincial process and creating a statement of joint principles (FMM 1997; Harrison 2007). These included a commitment to avoiding the imposition of 'unreasonable burden' on any region and to a collaborative federal-provincial effort based on comprehensive economic modelling (Harrison 2007). The JMM process was thus restarted under the rubric of the NCCP.

The NCCP featured regularized interaction among ministers and especially senior level officials, and extensive efforts at stakeholder engagement. However, the NCCP did not have the formal involvement of First Ministers and was characterized by a lack of agreement regarding purpose. Some participants, such as the Quebec government and Environment Canada, believed its mandate was to determine *how* to reach the target agreed to at Kyoto. Others, such as the Alberta government and Natural Resources Canada, believed the purpose was to determine *whether* that target could be reached. They also failed to address a basic issue – How would the total cost of reducing to 6 per cent below 1990 levels be divided amongst sources and provinces? All participants knew this was the fundamental matters, but were convinced agreement must first be reached on easier issues (confidential interviews).

The NCCP was marked by tension and outbursts of interjurisdictional conflict, with Ontario, Alberta and Quebec all walking out at various points. In 2001, after persuading the Kyoto parties to grant Canada credit for carbon sinks, the federal government announced its intent to ratify the Kyoto Protocol. This precipitated a major anti-ratification lobbying campaign, led by the oil industry and Alberta (Macdonald 2007). In the spring of 2002, as it became clear that the federal government planned to ratify Kyoto without federal-provincial agreement, Alberta withdrew from its position as co-chair of the NCCP, effectively ending the federal-provincial process (confidential interviews). The federal government refused provincial demands for an FMM and the federal Parliament ratified Kyoto unilaterally, over vociferous objections from all provinces (Harrison 2007). The federal government then developed its own program for regulating industrial emissions and provinces began moving forward independently with their own policy programs. By 2002 the federal-provincial process had broken down.

From 2003 to 2005, Prime Minister Paul Martin's Liberal government worked to develop a federal regulatory program, as well as bilateral agreements for emission reduction programs with willing provinces, using the inducement of federal funding. Six bilateral agreements were signed, setting out general provincial goals to be achieved with the aid of federal funding (Winfield and Macdonald 2008). In addition, the Martin government announced, in the *Canada Gazette,* plans to regulate industrial emissions using the Canadian Environmental Protection Act. Both initiatives ended with the defeat of the Liberal Martin government in late 2005.

From 2006 to 2008, the federal Conservative Harper government abandoned bilateral agreements, made no efforts to restart the JMM and took only desultory action to develop a federal program (Simpson et al. 2009). In 2008 following election of President Obama, the Harper government announced its plans to harmonize federal climate policy with that of the US. Thus Canadian and US national mitigation targets are now set at 17 per cent below 2005 levels by 2020. The federal government also adopted motor vehicle fuel efficiency standards that mirrored those adopted in the US. A 2011 analysis suggested that all the Canadian federal and provincial policies in place would only result in 25 per cent of the emissions cuts embodied in the national target for 2020.

By 2006 all Canadian provinces had formal climate change action programs, with targets and stated policies for meeting them (Marshall 2008). However there is little evidence of horizontal coordination between them in any forum of Canadian IGR. The exclusively provincial COF has discussed climate change at several annual meetings in an effort to develop common provincial policies but has failed to reach agreement, primarily due to objections by Alberta (Winfield and Macdonald, forthcoming). There is no coordination in relation to joint objectives or policy mechanisms and outcomes have been stuck at the lowest common denominator. The Council of the Federation's accomplishments are limited to issuing statements of ambiguous intent (COF 2007), reporting autonomous actions undertaken in various jurisdictions (COF 2007, 2008a) and committing to sharing information and best practices (COF 2008b). At the level of Ministers and bureaucrats, the Canadian Council of Ministers of Environment has not included climate change on its working agenda, even though this important and well established body of IGR has a well staffed permanent secretariat. In light of limited national horizontal coordination, some provinces, notably British Columbia, Manitoba, Ontario and Quebec, have joined cross-border arrangements such as the Western Climate Initiative, although the extent to which these commitments extend beyond rhetoric is open to debate.

ANALYSIS

This section returns to the analytic framework introduced above and applies it to our two cases in order to assess the extent to which, and along which dimensions, institutions of IGR differ.

Institutional Regularity and Predictability. In Australia, COAG was available as a forum over the whole period of study, although it was used intermittently. In Canada, the institutions of IGR over national climate policy coordination have been transient and unstable. The JMM/NCCP period from 1992 to 2002 provided a stable institutional context for intergovernmental interaction. However, the dissolution of these institutions and processes in 2003, and the fact that no stable institutional context dedicated to national climate policy coordination has existed from 2003 to the present, is a major point of difference from the Australian case and would appear to be a major barrier to the possibility of effective national coordination in Canada.

The institutions of IGR in Australia are somewhat paradoxical in terms of their predictability and regularity. The extra-constitutional standing of COAG has served to undermine the regularity of meetings, a situation that is exacerbated by the Prime Minister's control over convening and setting the agenda of COAG meetings. The implications of this are evident in the fluctuating frequency of COAG meetings depending on Commonwealth interest and ideology (Carroll and Head 2010). This is a distinct weakness of COAG and a source of ongoing calls for reform (BCA 2006).

In spite of this, there is considerable stability in COAG across time. The set of rules governing COAG regarding participation, agenda setting and decision making have persisted in spite of underlying shifts in political will and engagement and remain virtually unchanged since 1992. Furthermore, the COAG secretariat has existed as a site of institutional memory since its inception. When political actors began to reengage with efforts at national coordination in the 2005, efforts at coordination took place within COAG, first in a series of senior officials working groups and then in the formally mandated WGCCW and various sub-groups. At the horizontal level, the rules regarding agenda setting, participation and decision making within CAF have remained consistent since its inception in 2006.

Canadian institutions of IGR that have been used to coordinate national climate change policy are, in some ways, the inverse of those in Australia. The major institutions used to coordinate national policy in Canada have been fairly predictable when examined at particular points in time. The JMM featured regular meetings at both the ministerial and officials' levels and had stable participation from all jurisdictions. The JMM was organized along clear lines of authority, running from the Ministerial levels downwards and the NCCP recycled these institutional features, adding a permanent secretariat and augmenting the density and institutional stability of interjurisdictional interaction through creation of the NAICC-CC and a set of seven working groups and 16 issue tables.

This regularity was undercut by instability in the face of underlying changes in political will and engagement. The JMM was damaged in 1997, following the decision by the Chrétien government to disregard federal-provincial agreement over the Kyoto negotiations. The NCCP created in the wake of Kyoto ceased to exist following Alberta's departure in 2002. Since 2003 there has been no institution of IGR to facilitate coordination of national climate change policy and policy action has devolved into a set of unilateral initiatives and a series of nascent efforts at cross-border coordination. This instability across time, in concert with non-use of the FMM, is amplified by the absence of any meaningful linkages between efforts at coordination at ministerial and official level and with the First Ministers, who possess the ultimate decision making authority in Canadian politics, given the centralization of power in the offices of the Prime Minister or Premier (Savoie 1999).

Participation. The participation of decision makers in the process is a key difference between our cases. The JMM/NCCP in Canada suffered from the absence of First Minister engagement. As a result these processes were unable to resolve underlying tensions between the two lead ministries from each government (energy and environment) and lacked political impetus and decision making authority. The absence of First Minister engagement with the process, as evidenced by the fact that the FMM never once convened explicitly to discuss national climate policy, is surely one of the core weaknesses of Canadian efforts at intergovernmental coordination (Harrison 2007). For example the federal government was able, under the rules governing the convening of an FMM, to ignore the call by all provinces for a FMM to discuss Kyoto ratification in 2002.

The Council of Australian Governments on the other hand, provided an institutional venue for a process of national climate policy coordination that directly involved Australian First Ministers. This participation rule helped ensure that, when the Commonwealth government began to once again engage in national climate policy discussions with the states/territories in 2005, the process involved the ultimate bearers of decision making authority. Following the electoral shift of 2007, COAG was immediately convened and climate change placed at the top of the agenda. The WGCCW was created, signifying a centralization of the issue – taking it away from the ministerial councils and placing it under the direct supervision of the First Ministers (Jones 2010). The various subgroups operating under the WGCCW worked to foster agreement and coordinate national positions amongst jurisdictions, although the extent to which these reflected true coordination on the imposition of national interests is open to some challenge (ibid).

The Council of the Australian Federation similarly brought together First Ministers endowed with the authority to make decisions and reach agreement on coordinated positions on climate change – a process of deal making evident during the 18-month period of negotiations that led to agreement on NETS (Tiernan 2008). More importantly, the relationship between CAF and COAG appears to have created positive tension that helped to spur national coordination on climate policy. The ability of subnational jurisdictions to reach agreement on national objectives and policy initiatives (NETS 2007; CAF 2007a) appears to have contributed to the Commonwealth decision to reengage in efforts to develop nationally coordinated policy in the mid-2000s (Tiernan 2008).[3]

Codification, Specification & Enforcement. In terms of specificity, codification and enforceability of commitments, there is little difference between our two cases. In Australia, both COAG and CAF have produced agreements related to climate change that are primarily informal and uncodified. The National Emissions Trading Scheme, as well as the national strategies on energy efficiency and Carbon Capture and Storage (CCS), were joint communiqués and strategic documents, rather than formal Intergovernmental Agreements allocating responsibilities and setting out funding arrangements. Those agreements that have been formally codified – the IGAs creating COAG and CAF – suffer from inherent weakness, as they are neither legally binding nor judicially actionable, as is also the case for Canadian IGAs.

Despite the paucity of formal, legal binding agreements in Australia, making interjurisdictional coordination initiatives public may increase the credibility of commitments. The national strategy on energy efficiency is set forth in a formal document. Principles to guide reconciliation of Commonwealth and state policies have been made public. Modelling on the impacts of national policies has been conducted and released. Funding programs have been developed by the Commonwealth government (the Climate Change Action Fund, for example) to gather and redistribute carbon pricing revenues and to facilitate energy efficiency and CCS programs in states/territories (COAG 2009c; COAG 2010; Access Economics 2009; Commonwealth of Australia 2008).

Enforcement is an ongoing problem in Canadian IGR, as lines of accountability and legal authority run from government to the electorate, rather than from government to government. In terms of climate policy there has been little to enforce, since agreement has not been reached. The exception is the joint agreement on national targets reached at the JMM in Regina prior to the Kyoto COP3. That this agreement was subsequently disregarded by the federal Chrétien government demonstrates the lack of credible commitment embodied in the process and an inability to enforce agreements, and illustrates a powerful source of mistrust between the federal and provincial governments.

Decision rules. Decision rules do not seem to be significant in national climate policy coordination in our cases. In all institutions participants have a *de facto* veto. The Canadian JMM/NCCP relied on consensus, although this was never codified or explicitly agreed to but rather seems to have been imported from the broader system of IGR in Canada. In Australia, COAG similarly relied on consensus and, despite the use of qualified majority voting in some of the ministerial councils, there is little evidence to suggest that this affected intergovernmental coordination. In Australia, both CAF and COF operate on the basis of consensus and there is no discernible difference between these institutions in terms of decision making.

Bureaucracy. With regard to bureaucratic mandate and resources, the differences between our cases are subtle. In Australia, COAG and CAF have dedicated and permanent secretariats with an explicit mandate to foster coordination between jurisdictions. The extent to which these are autonomous, however, is limited by both funding and location arrangements, with the COAG secretariat located within the Commonwealth Department of the Prime Minister and the CAF secretariat located within the jurisdiction of the rotating chair. These arrangements likely limit the secretariats' capacity for autonomous action in achieving intergovernmental coordination. This lack of autonomy is reflected and reinforced, in the funding arrangements for each, with the Commonwealth providing the bulk of funding for the COAG secretariat and the rotating chair funding CAF.

Looking to Canada, the most significant difference is in regards to stability across time. The JMM process of 1992–2002 was supported by a federal-provincial secretariat with a clear mandate and significant resources. However, this body ceased to exist in 2002 and no secretariat for Canadian climate policy has emerged in its place. Thus the permanence of the national secretariat was undercut by the impermanence of the intergovernmental process itself, and the former was officially disbanded following the cessation of the latter. This is further underscored by the absence of a permanent secretariat to support FMM in Canada.

CONCLUSION

Does the successful coordination of Australian climate policy, as compared with the Canadian experience, point to the importance of particular modes of coordination or institutions of IGR? Assessing the success of economic reform in Australia in areas of joint jurisdictional responsibility, Braun (2006) points to the importance of reforms to the institutions of Australian federalism undertaken in the early 1990s. Australia expanded on the predominantly adversarial system of IGR by creating new mechanisms and institutions of IGR (i.e., COAG) based on voluntary engagement, bargaining and coordination. The result according to Braun (2006), is that Australia now runs 'on two engines when dealing with concurrent policy areas' with the second engine, collaboration, institutionalized as part of the regular process of Australian IGR.[4] Does this help to explain success and failure in the case of national climate policy coordination?

Our analysis suggests a relationship does exist between national policy coordination and institutional strength and that the latter must be considered when explaining the difference in coordination outcomes in our two cases. In line with findings in the extant literature on institutional strength we find that the institutions of IGR used to coordinate climate policy in Australia are stronger than their Canadian counterparts (Bolleyer 2009; Brown 2003). The stability and predictability of COAG is contrasted with the transience of the JMM/NCCP in Canada (and the inherent weakness and instability of the FMM). The rules for work practices, the existence of a supportive bureaucratic body, and the inclusion of First Ministers in COAG all provided an opportunity for high level interaction that has had no parallel in the Canadian process. Both institutional contexts suffer from the unpredictability resulting from the concentration of power in the central government, reliance on consensus decision making rules and weak enforcement mechanisms, but the Australian institutions of IGR that were used to coordinate national climate policy appear to have provided an enabling context for effective coordination when the requisite political will emerged. That context that does not appear to exist in Canada.

This leaves the question as to how these differences in institutional strength translate into differences in coordinative capacity. Time constraints limit the extent to which we can engage this important question and so we offer only a preliminary conjecture that requires further consideration in future research. Braun (2006) makes the suggestion that successful collaborative coordination in economic policy in Australia depended on assertive central leadership through deployment of political as well as fiscal levers of influence. This sort of aggressive action by the central government was essential as a means of expanding win-sets so as to create overlapping areas of agreement amongst the constituent jurisdictional actors, and seems to fit well with the assertiveness of the Commonwealth in centralizing the process within COAG and the WGCCW in 2007 and after. Note that this argument, while similar to those who advocate increased centralization and federal assertiveness as the most effective means of achieving national environmental policy objectives (Weibust 2009), differs in that it remains predicated on interjurisdictional cooperation rather than the assertion of federal fiat.

Braun (2006) acknowledges this explicitly, noting that assertions of central leadership come with considerable potential problems, in that they lack built-in mechanisms of constraint or self control. These are essential in order to maintain the delicate balance between autonomy and interdependence; in Canada perhaps even more so than in Australia, due to the decentralizing forces that exist in the former. Central domination may be no more the answer to the coordination challenge than decentralized competition – the answer lies neither in Hobbes nor Hayek. Indeed this is where the institutional stability of COAG in Australia may increase coordinative capacity, by creating the practices and norms of cooperation and reducing temptations to cheat that have plagued Canadian efforts at national climate policy coordination, while maintaining space for jurisdictional innovation and responsiveness to local considerations. COAG has served, perhaps, as the means through which a jurisdictional voice can be combined with central assertiveness and leadership. However, the Australian experience presents a cautionary tale. In spite of the successful passage of national carbon pricing legislation, there has been increasing public pushback and criticism from Australian states (those controlled by Labor as well as by the Liberals) regarding the lack of consultation and over-centralization of the process (Salusinszky 2011a, b), pointing to the danger of exit strategies should opportunities for an effective voice remain stifled (Hirschman 1973). In addition there is a risk that coordinative agreements could unravel should the political will or interest of the Commonwealth government decline. Further research is needed in order to better understand the balance between centralization and collaboration in the intergovernmental process and the institutional factors that may enable successful balancing of these competing imperatives.

There are noticeable and meaningful differences between the institutions of IGR used for national climate policy coordination in our cases. As noted above, we do not contend that these differences fully explain divergent outcomes in the two countries – there are numerous factors that contribute to this outcome as outlined above. However, we do contend that the institutional differences outlined in this chapter contribute to a fuller and more nuanced understanding of the challenges and constraints that have limited national climate policy coordination in Canada, and have enabled such coordination in Australia. Our analysis suggests that specific institutional weaknesses in Canadian IGR, participation rules and temporal stability, rather than decision making or enforcement mechanisms, may have contributed to the inability to effectively coordinate between jurisdictions. While recognizing the 'institutional conservatism' of IGR in Canada (Simeon 2006), we nonetheless suggest that this points to the need for creative thinking and incremental changes to this institutional context, as a way to open up pathways to future progress.

NOTES

1. This pressure runs in both directions, as recent moves in the US to increase vehicle emissions standards led to harmonization efforts in Canada.
2. Author's personal communication with David Oulton, 16 November 2010.

3. The arrow likely runs in the opposite direction as well, as the credible threat of unilateral federal imposition resulting from the considerable asymmetry between subnational and Commonwealth government in Australia creates substantial incentives for states/territories to develop coordinated positions in order to maximize their ability to resist central imposition and influence national policy.
4. Braun (2006) acknowledges that the first engine – adversarial/competitive intergovernmental relations – coexists with this second, collaborative, mode.

REFERENCES

Access Economics Pty Limited (2009), *Report 1: Impact of CPRS – Fiscal Report*. Retrieved on 9 September 2011, from: http://www.caf.gov.au/policyinnovation.aspx.

Ayers, C. (2006), 'Australian Intergovernmental Relations and the National Emissions Trading Scheme,' *Melbourne Journal of Politics* **31**, 36–55.

Berkovic, N. and C. Hart (2008), 'States Revolt over Rudd's Carbon Plan', *The Australian*, 14 November, 2008.

Biermann, F. (2001), 'The Emerging Debate on the Need for a World Environmental Organization: A Commentary,' *Global Environmental Politics* **1**(1), 45–55.

Bolleyer, Nicole (2009), *Intergovernmental Cooperation – Rational Choices in Federal Systems and Beyond*. Oxford: Oxford University Press.

Bolleyer, N. and T. Börzel (2010), 'Non-Hierarchical Policy Coordination in Multi-level Systems,' *European Political Science Review* **2**(2), 157–85.

Bolleyer, N. and E. Bytzek (2009), 'Government Congruence and Intergovernmental Relations in Federal Systems,' *Regional and Federal Studies* **19**(3), 371–97.

Braun, D. (2006), 'Between Market Preserving Federalism and Intergovernmental Coordination: The Case of Australia,' *Swiss Political Science Review* **12**(2), 1–36.

Brown, Douglas (2003), *Market Rules: Economic Union Reform and Intergovernmental Policy-Making in Australia and Canada,* Montreal: McGill-Queen's University Press.

Cameron, D. and R. Simeon (2002), 'Intergovernmental Relations in Canada: The Emergence of Collaborative Federalism,' *Publius* **32**(2), 49–71.

Carroll, P. and B. Head (2009), 'The Organization and Management of Intergovernmental Relations in Australia,' Working paper available at: www.psa.ac.uk/journals/pdf/5/2010/127_805.pdf.

Christoff, P. (2005), 'Policy Autism or Double-Edged Dismissiveness? Australia's Climate Policy under the Howard Government,' *Global Change, Peace & Security* **17**(1), 29–44.

Commonwealth of Australia (2008), *Australia's Low Pollution Future: The Economics of Climate Change Mitigation*. Retrieved on 8 September 2011, from: http://www.treasury.gov.au/lowpollutionfuture/report/html/01_Chapter1.asp.

Commonwealth of Australia (2011), *Securing a Clean Energy Future: The Australian Government's Climate Change Plan*. Retrieved 10 October 2011 from: http://www.cleanenergyfuture.gov.au/clean-energy-future/our-plan/.

Council of the Australian Federation (CAF) (2006), *CAF Communique – October 2006*. Retrieved 8 September 2011, from: http://www.caf.gov.ca/meetings.aspx.

Council of the Australian Federation (CAF) (2007a), *CAF Communique – April 2007*. Retrieved 10 September 2011, from: http://www.caf.gov.au/meetings.aspx.

Council of the Australian Federation (CAF) (2007b), *Council of the Australian Federation Declaration on Climate Change*. Retrieved 10 September 2011, from: http://www.caf.gov.au/policyinnovation.aspx.

Council of Australian Governments (COAG) (2005), *Communique June 2005*. Retrieved 9 September 2011, from: http://www.coag.gov.au/coag_meeting_outcomes/archive.cfm.

Council of Australian Governments (COAG) (2007a), *Communique December 2007*. Retrieved 7 September 2011, from: http://www.coag.gov.au/coag_meeting_outcomes/archive.cfm.

Council of Australian Governments (COAG) (2007b), *Communique November 2007*. Retrieved 6 September 2011, from: http://www.coag.gov.au/coag_meeting_outcomes/archive.cfm.

Council of Australian Governments (COAG) (2008a), *Communique March 2008*. Retrieved 10 September from: http://www.coag.gov.au/coag_meeting_outcomes/archive.cfm.

Council of Australian Governments (COAG) (2008b), *Communique October 2008*. Retrieved 8 September 2011 from: http://www.coag.gov.au/coag_meeting_outcomes/archive.cfm.

Council of Australian Governments (COAG) (2009a), *Communique July 2009*. Retrieved 6 September 2011 from: http://www.coag.gov.au/coag_meeting_outcomes/archive.cfm.

Council of Australian Governments (COAG) (2009b), *Communique April 2009*. Retrieved on 8 September 2011, from: http://www.coag.gov.au/coag_meeting_outcomes/archive.cfm.

Council of Australian Governments (COAG) (2006), *Council of Australian Governments' Plan for Collaborative Action on Climate Change – Attachment C*. Retrieved 6 September 2011 from: http://www.coag.gov.au/coag_meeting_outcomes/2006-02-10/docs/attachment_c_climate_change.pdf.

Council of Australian Governments (COAG) (2010), 'Final Report: Jurisdictions' Reviews of Existing Climate Change Mitigation Measures.' Report submitted by the Complementary Measures Sub-Group to the COAG Senior Officials Working Group.

Council of Australian Governments (COAG) (2009), *National Strategy on Energy Efficiency*. Retrieved 8 September 2011, from: http://www.coag.gov.au/reports/docs/nsee_update_july_2010.rtf.

Council of the Federation (COF) (2008a), *Climate Change: Fulfilling Council of the Federation Commitments*. Retrieved 5 September 2011 from: http://www.counciloffthefederation.ca/meetings/meetings2008.html.

Council of the Federation (COF) (2008b), *Canada's Premiers: Taking Action on Climate Change.* Retrieved 6 September 2011 from: http://www.councilofthefederation.ca/meetings/meetings2008.html.
Council of the Federation (COF) (2007), *Climate Change: Leading Practices by Provincial and Territorial Governments in Canada.* Retrieved 5 September 2011 from: http://www.councilofthefederation.ca/meetings/meetings2008.html.
Crowley, K. (2007), 'Is Australia Faking It? The Kyoto Protocol and the Greenhouse Policy Challenge,' *Global Environmental Politics* **7**(4), 118–39.
Elazar, Daniel (1987), *Exploring Federalism,* Tuscaloosa: University of Alabama Press.
First Minister's Meeting (FMM) (1997), *Joint Communique – 12 December 1997.* Retrieved 5 September 2011, from: http://www.scics.gc.ca/cinfo/80003606_e.html.
Hardin, G. (1968), 'The Tragedy of the Commons', *Science* 162(3859): 1243–8.
Harris, P. (2007), 'Collective Action on Climate Change: The Logic of Regime Failure,' *Natural Resources Journal* **47**, 195–224.
Harrison, K. (2007), 'The Road Not Taken: Climate Change Policy in Canada and the United States,' *Global Environmental Politics* **7**(4), 92–117.
Harrison, K. and L. Sundstrom (2007), 'The Comparative Politics of Climate Change,' *Global Environmental Politics* **7**(4), 1–18.
Hirschman, Albert (1973), *Exit, Voice, and Loyalty: Responses to Decline in Firms, Organizations, and States*, Cambridge: Harvard University Press.
Hollander, R. and H. Patapan (2007), 'Pragmatic Federalism: Australian Federalism from Hawke to Howard,' *The Australian Journal of Public Administration* **66** (3), 280–97.
Hurst, D. (2011), 'Carbon Tax no Calamity for Queensland, says Report,' *Brisbane Times*. 23 August 2011.
Johnston, M. (2011), 'Carbon Tax Bills Pass Lower House of Federal Parliament,' *Perth Now*, 20 October 2011.
Jones, S. (2010), 'The National Renewable Energy Target: An Example of Post-Machiavellian Policy-Making?' *Australian Journal of Public Administration* **69**(2), 165–77.
Keating, Michael and John Wanna (2000), 'Remaking Federalism', in Michael Keating and John Wanna (eds), *Institutions on the Edge? Capacity for Governanc*e, St Leonards, Crows Nest, East Melbourne: Allen & Unwin.
Kenny, M. (2011), 'Premiers can't Persuade PM on Carbon Tax,' *Adelaide Now,* 20 August 2011.
Lyster, R. (2004), 'Common but Differentiated? Australia's Response to Global Climate Change,' *The Georgetown International Environmental Law Review* **16** (561), 561–91.
McKenna, M. and I. Salusinszky (2011), 'States Step Up Carbon Tax Compo Fight,' *The Australian*, 19 August 2011.
Macdonald, Douglas (2007), *Business and Environmental Politics in Canada,* Peterborough: Broadview Press.
Marshall, Dale (2008), *Provincial Power Play: Breaking Away from Federal Inaction on Climate Change*, Vancouver: The David Suzuki Foundation.
Meekison, Peter J., Hamish Telford and Harvey Lazar (2004), 'The Institutions of Executive Federalism: Myths and Realities,' in Peter Meekison, Hamish

Telford and Harvey Lazar (eds), *Canada: The State of the Federation, 2002, Reconsidering the Institutions of Canadian Federalism,* Montreal: McGill-Queens Press.
Montpetit, Eric (2003), *Misplaced Distrust: Policy Networks and the Environment in France, the United States, and Canada*, Vancouver: UBC Press.
National Emissions Trading Taskforce (NETT) (2006), *Possible Design for a National Greenhouse Gas Emissions Trading Scheme.* Retrieved 4 September 2011 from: http://www.climatechange.gov.au/government/initiatives/cprs/~/media/publications/cprs/NETT-Final.ashx.
North, Douglass (1990), *Institutions, Institutional Change, and Economic Performance*, New York: Cambridge University Press.
Olson, Mancur (1965), *Logic of Collective Action; Public Goods and the Theory of Groups,* Cambridge: Harvard University Press.
Ostrom, Elinor (1990), *Governing the Commons: The Evolution of Institutions of Collective Action*, Cambridge: Cambridge University Press.
Packham, B. (2011), 'Four Biggest States Rebuffed on Carbon Tax Compensation,' *Perth Now*, 19 August 2011.
Painter, M. (1998), *Collaborative Federalism: Economic Reform in Australia in the 1990s*, Cambridge: Cambridge University Press.
Painter, M. (1991), Intergovernmental Relations in Canada: An Institutional Analysis *Canadian Journal of Political Science* **24** (2), 269–88.
Painter, M. (2001), 'Multi-level Governance and the Emergence of Collaborative Federal Institutions in Australia,' *Policy & Politics* **29** (2), 137–50.
Papillon, Martin and Richard Simeon (2004), 'The Weakest Link? First Ministers Conferences in Canadian Intergovernmental Relations', in Peter Meekison, Hamish Telford and Harvey Lazar (eds), *Canada: The State of the Federation 2002, Reconsidering the Institutions of Canadian Federalism*, Montreal and Kingston: McGill-Queen's University Press.
Salusinszky, I. (2011), 'COAG's Love-in All Over Now,' *The Australian,* 13 August 2011.
Salusinszky, I. and R. Barrett (2011), 'Tax to Join Agenda at Reset COAG meeting,' *The Australian*, 27 July 2011.
Savoie, Donald (1999), *Governing from the Centre: The Concentration of Power in Canadian Politics*, Toronto: University of Toronto Press.
Scharpf, F. (1988), 'The Joint-Decision Trap: Lessons from German Federalism and European Integration,' *Public Administration* **66** (3), 239–78.
Selin, Henrik and Stacy D. VanDeveer (eds) (2009), *Changing Climates in North American Politics: Institutions, Policymaking and Multilevel Governance*, Cambridge: MIT Press.
Simeon, Richard (2006), *Federal-Provincial Diplomacy: The Making of Recent Policy in Canada*, Toronto: University of Toronto Press.
Simeon, Richard and Amy Nugent (2008), 'Parliamentary Canada and Intergovernmental Canada: Exploring the Tensions', in Herman Bakvis and Grace Skogstad (eds), *Canadian Federalism: Performance, Effectiveness and Legitimacy*, Don Mills: Oxford University Press Canada.
Simpson, Jeffrey, Mark Jaccard and Nic Rivers (2007), *Hot Air: Meeting Canada's Climate Change Challenge,* Toronto: McClelland and Stewart.

Stilborn, J. (2003), 'Canadian Intergovernmental Relations and the Kyoto Protocol: What Happened, What Didn't.' Conference paper retrieved from: http://www.cpsa-acsp.ca/paper-2003/stilborn.pdf.

Strebel, F. (2011), 'Inter-governmental Institutions as Promoters of Energy Policy Diffusion in a Federal Setting,' *Energy Policy* **39**, 467–76.

Taylor, L. (2008), 'Rudd cools on Carbon Targets,' *The Australian*, 5 December 2011.

Thorlakson, L. (2003), 'Comparing Federal Institutions: Power and Representation in Six Federations,' *West European Politics* **26** (2), 1–22.

Tiernan, A. (2008), 'The Council for the Australian Federation: A New Structure of Australian Federalism', *The Australian Journal of Public Administration* **67**(2), 122–34.

Wade, M. (2011), 'Carbon Tax Job Outlook Upbeat,' 20 August 2011.

Weibust, Inger (2009), *Green Leviathan,* Toronto: Ashgate.

Winfield, Mark and Douglas Macdonald (2008), 'The Harmonization Accord and Climate Change Policy: Two Case Studies in Federal-Provincial Environmental Policy,' in Grace Skogstad and Herman Bakvis (eds), *Canadian Federalism 2nd ed*, Don Mills: Oxford University Press Canada.

Winfield, Mark and Douglas Macdonald (forthcoming), 'Federalism and Canadian Climate Change Policy,' in Grace Skogstad and Herman Bakvis (eds), *Canadian Federalism 3rd ed.*, Don Mills: Oxford University Press Canada.

Vasek, L. (2011), 'NSW Puts Carbon Tax Offer to Canberra,' *The Australian*, 7 September 2011.

9. Allocating greenhouse gas emission reductions amongst sectors and jurisdictions in federated systems: the European Union, Germany and Canada

Douglas Macdonald

Given differing per-unit reduction costs and a desire to maximize efficiency, a basic challenge facing climate change mitigation policy makers is allocating the total reduction effort amongst groups of sources of greenhouse gas (GHG) emissions, such as transportation, buildings or different industrial sectors. In a unitary state, while having implications for local or regional economies, the focus is upon sectoral allocation. At the international level, on the other hand, attention is focused upon allocation amongst states, rather than sources. This is because normative concerns, related to historic responsibility for the problem and differing capacity to act, have led to agreement that Northern states should reduce more than Southern states. It is also because, at the international level, all states face the problem of collective action – the need to guard against free-riding – and so seek international agreements ensuring that all will act. The major such agreement to date, the 1997 Kyoto Protocol, allocated a total reduction amongst the Northern states.

In federated systems,[1] there is a need to consider allocation amongst *both* sources and 'states', in this case the subnational jurisdictions in which sources are located. This is because, for the efficiency reasons noted above, sources in one region may have to reduce more than those in another. Even if there is no conscious intent to allocate geographically amongst subnational jurisdictions, the mere fact of allocating amongst sources inevitably has that effect. This is not a political problem when sources are distributed roughly evenly throughout the federated system, since the costs borne by the subnational jurisdictions are roughly equal. It is, however, a problem when the sources being asked to significantly reduce are geographically concentrated and important to the economy of the subnational jurisdiction in which they exist. In that case, the subnational government sees that its jurisdiction is being asked to bear a greater portion of the total cost than other jurisdictions in the federated system. For that reason, understandably, that government may resist efforts to develop climate change policy. Examples of geographically concentrated sources in federated systems, which are important to the economy of the relevant subnational jurisdiction, include coal mining in Australian and US states, the oil and gas industry in western Canada, automobile manufacturing, which tends to be concentrated in countries such as the US, Canada and Germany, coal-produced electricity, and the coal industries of Germany and other parts of the EU.

Subnational and sectoral allocation can be done automatically through the market mechanisms of a carbon tax or trading system. However, the design of such systems has geographic equity implications and so may bring into play subnational government resistance.

Where the national government has a clear constitutional mandate to implement climate change policy, it might be argued it could do so unilaterally, ignoring its subnational governments. While true in a legal sense, this is difficult in a political sense since the national government is seeking support from voters living within those subnational jurisdictions, whose views are influenced by the actions and words of their subnational government and of economically important sources. For that reason, the national government is susceptible to lobbying influence by both the relevant subnational government and its concentrated sources. As noted, even if such a national government is attempting to allocate reductions only amongst sources, those decisions have regional implications, which means allocation amongst subnational jurisdictions will inevitably be a factor in its climate policy decision making. Thus, in federated systems, to achieve effective policy there is a need to consider allocation amongst both sources and subnational jurisdictions in order to address potential resistance by relevant subnational governments.

This basic challenge which faces climate change policy makers within federated systems, is the subject of this chapter. Because climate policy has been relatively ineffectual to date, in comparison to such things as the political activity of the fossil-fuel industry or the sheer complexity of the 'wicked problem,' subnational government resistance has not been the most important factor slowing the pace of climate change policy development. However, if climate policy is to achieve anything like the GHG reductions called for by science, it must impose much higher short-term costs upon sources. Countries like Australia, should it proceed with national policy, and the US, which has yet to do so, will be forced to grapple with the problem and those which have already done so, such as the EU, will again confront the issue, but on a much larger scale. It is essential that we understand the basic dynamic. The purpose of this chapter, accordingly, is to identify factors which influence the ability of a federated system to reach agreement on cost allocation amongst subnational jurisdictions and thereby remove a potential roadblock to effective climate change policy.

Three federated systems are examined: the EU, Germany and Canada. The major subnational governments examined in each are the energy-intensive jurisdictions of Poland and other eastern states in the EU, North Rhine Westphalia in Germany, and Alberta and Saskatchewan in Canada. Examining these three allows comparison of a case in which allocation amongst subnational governments[2] was explicitly addressed and successfully agreed upon (the EU, in both 1997, 1998 and again in 2008); a case in which such allocation was implicit and also successful (Germany), and a case (Canada) in which such allocation, and indeed any form of co-ordinated policy, has failed. Why were the EU and Germany successful, while Canada was not?

The next section sets out the variables examined, followed by the three case studies. Each provides a chronological review, followed by discussion of the way in which the three variables under examination help to explain the outcome in that case. The conclusion attempts to answer the question posed above and based on that provides suggestions for ways in which federated systems can best address the issue.

INSTITUTIONS AND THE ROLES OF SUBNATIONAL AND NATIONAL GOVERNMENTS

In the case study analysis which follows, the focus is upon the factors most relevant to the ability of a subnational government to resist federation-wide policy, whether implemented by the national government or by both levels of government co-ordinating policy through an intergovernmental relations (IGR) process. Three are considered most important: 1) those aspects of the institutional make-up of the federated system which influence the way in which subnational governments participate in climate policy processes; 2) within those parameters, the role of potentially resisting subnational governments; and 3) the role of the national government. Each is briefly discussed.

Federal Institutions

Institutions are defined here as the rules, formal and informal, which determine the ways in which governments make climate change policy decisions in federated systems. Bakvis and Skogstad (2012, 3) set out three sets of institutions relevant to federalism: 'the constitutional division of powers … the institutions of intrastate federalism that provide for the representation of constituent units within the central government … and third, the institutions and processes of interstate federalism through which the two orders relate directly to one another.' Using the Bakvis and Skogstad typology, three institutions are examined: 1) those which determine jurisdiction of the subnational government respecting climate policy; 2) if policy is made by the national government, those which determine the ability of subnational governments to directly participate; and 3) if it is made by IGR, those which determine whether decisions are made by consensus or majority vote.

With respect to the first, the assumption is that a subnational government will have greater influence if it is able to regulate GHG sources located within its borders. Even if jurisdiction is shared, and the national government is also regulating, subnational government jurisdiction grants constitutional power. Although space precludes full discussion in the cases, this variable is complicated by the fact that climate policy encompasses many different policy activities, including transportation policy, regulation of land-use, building codes, environmental regulation, natural resource extraction and energy policy. Subnational governments may hold jurisdiction in some of these areas and not others.

The second and third categories relate to a central institutional question concerning climate change policy in federated systems. Is it made by the national government alone, or by both levels of government jointly? If the former, the issue is subnational government's ability to influence national government policy. Is it by direct participation inside that process or only from outside, by lobbying? An issue related to the former is territorial representation in the upper house. Is it the citizens of the subnational government who are represented, as in the US Senate, or the subnational government itself, as in the German Bundesrat (Hueglin and Fenna, 2006: 60)? The latter system, obviously, gives greater power to the subnational government.

Third, the form of IGR decision making is relevant, since consensual decision making tends towards lowest common denominator decisions, which will favour the resisting subnational government. As Member States have increasingly ceded power to the EU level, that body has moved from consensus to qualified majority voting. However, as discussed below, EU climate cost allocation has been done to date completely by consensus. Others aspects of the IGR process, such as predictability, regularity of meetings and specificity of intergovernmental agreements may also be relevant (Bolleyer, 2009 – see the discussion in Chapter 8, Gordon and Macdonald). For these purposes, however, the subject is limited to the one question of consensus relative to qualified majority or absolute majority voting.

The Role of the Subnational Government

The ability of a subnational governments to resist is determined by the political power it can wield in the federated system, beyond that conferred by the institutional context. Galbraith (1983) argues that the power of an actor flows from its available resources, degree of organization and degree of motivation. In terms of the last, while many factors, such as subnational government ideology or lobbying by environmentalists will influence degree of motivation, here we focus upon only one – motivation stemming from the fact the subnational jurisdiction is being asked to pay a higher cost than others. In the cases below, that cost differential is represented by differences in per capita emissions, based upon the assumption that higher per capita emissions indicate the greater reliance of a subnational economy on certain activities, such as oil extraction or electricity generation by coal, which if curtailed will have economic impacts.

By definition, the subnational government is organized (at least in comparison to an actor such as a social movement), which is part of the unique challenge of cost allocation in federated systems: in them, unlike in unitary states, regional non-state actors are represented by an organized state. Furthermore, again by dint of being a government, that actor will have financial resources secured by tax revenue. The *degree* of such resources is measured in the cases below by the per capita GDP of the subnational jurisdiction. Finally, we need to consider the size of the subnational jurisdiction economy relative to that of the economy as a whole. The larger that is, the greater will be its share of total emissions and the less easily can it simply be ignored by other governments without sacrificing policy effectiveness, thus increasing its influence. In the cases below, comparative data is provided on subnational jurisdictions' share of total emissions.

The Role of the National Government

Kelemen (2004) argues that the nature of the national government influences the degree of centralization in a federated system – the more fragmented (such as the US checks and balances system) the more it will feel a need to control subnational governments. Other aspects of the internal workings of the national government, in particular the relative power of environment and energy departments, are pertinent to its role in climate policy. Space precludes consideration of such factors in the case studies and instead the focus is upon two things – national government motivation and the strategy it uses to address the need for subnational jurisdiction cost allocation.

The former is important because the more motivated a national government is to put in place effective climate change policy – the more likely it is to exert countervailing power to influence the resisting subnational government. Beyond national government motivation flowing from governing party ideology or lobbying by business and environmentalists, two factors differentiate it from subnational governments. The first is the fact that concentrated sources are less of a problem for national than subnational governments – costs imposed upon concentrated sources will have greater impact upon the subnational than the national economy, meaning those sources are better able to influence the subnational than the national government. As Weibust (2009, 201) puts it: 'If a large employer threatens exit, a larger jurisdiction can better afford to call the firm's bluff.' Second, external factors are likely to be more important for the national than subnational governments, since the former has greater responsibility for foreign policy. In the cases, we see that a desire for global leadership prompted the EU to want effective policy while in Canada a desire to harmonize with the US led to the opposite.

As well as motivation, the other variable that is significant in this comparison is the strategy used by the national government to bring about agreement on cost allocation. This includes various means of threatening and cajoling subnational governments, but most important is how it addresses the allocation issue. Should it be ignored or explicitly addressed? The question is important because it is difficult to use side-payments to broker agreement if the issue is ignored or only addressed implicitly. Weibust (2009, 203) notes that the 'capacity to offer side-payments helps overcome distributional issues that can be a barrier to agreement.' The first element in national government capacity is the decision to explicitly address subnational jurisdiction allocation. As we shall see, Germany and Canada have opted for implicit allocation, and the EU for explicit. Since the same strategy is associated with German success and Canadian failure, the significance of this variable needs to be examined with care.

To summarize, the hypothesis offered here is as follows. First, a subnational government seeking to shield its concentrated sources from reduction costs will best be able to do so if it has jurisdiction, can directly participate in national government decision-making (when decisions are made in that forum) or, when by IGR, are made only by consensus. Second, it is assumed it will best be able to do so if it is highly motivated because of higher reduction costs relative to others in the federated system, and is a larger, richer subnational jurisdiction. Thirdly, it is assumed subnational government resistance will be diminished if a motivated national government exerts countervailing power and is adept in strategies used to bring about agreement on cost allocation, including when necessary explicit allocation with the use of side-payments.

CLIMATE CHANGE POLICY IN THE EU

The EU has twice allocated an overall emission reduction goal amongst its Member States, both times during the run-up period to an international UNFCCC meeting at which it hoped to play a lead role, knowing that it could only do so if it had first made a credible commitment to domestic action within Europe – the 1997 Kyoto Conference of Parties (CoP) to the UNFCCC and the 2009 Copenhagen Conference of Parties were the occasions. Prior to 1997, the climate regime parties agreed that the EU could participate in UNFCCC negotiations on the basis of a 'bubble' commitment, applied collectively to all Member States. The strategy used by the EU to press for action at the international level was to gain credibility by adopting a stringent domestic target and then advocating something similar for the international regime – however, as Vogler explains (2011, 151): 'these declaratory targets, to be credible, required an allocation of burdens (or "effort") amongst the Member States.' In 1990, the Commission produced a draft climate policy program, including the concept of differentiated Member State targets, based on experience in negotiating acid rain targets which had culminated, after difficult negotiations, in an agreement in 1988 (Jordan and Rayner 2010). However, such a top-down approach led by the Commission did not have the necessary support in the Council. Aware that EU leadership at the upcoming Kyoto COP depended upon prior agreement about allocation, the Commission twice proposed allocations in the fall of 1996, both of which were rejected by Member States. Then, beginning in January 1997, the Dutch Council Presidency asked outside experts at Utrecht University to develop a proposal for allocation amongst Member States, based upon a technical analysis (referred to as the Triptych Approach) of possible emission reductions in three sectors: (1) transport and housing; (2) energy; and (3) heavy industry, combined with national circumstances (Cass 2006; Lacosta et al 2007; Haug and Jordan 1010). This new approach, coupled with pre-existing Member State commitments and pressure from the need to agree prior to the December Kyoto meeting, allowed agreement on allocation to be negotiated at the Council meeting of March, 1997 – but only for a total reduction of 10 per cent not the 15 per cent which the Dutch Presidency wanted to take into the Kyoto negotiations. This was because Germany, Belgium, the UK, Greece and Portugal rejected the allocation recommended by the Dutch Presidency, totalling 15 per cent, because the Netherlands had only given itself a 5 per cent reduction (Vogler 2011, 158). To get around this stalemate, it was agreed the EU would advocate a 15 per cent reduction, but only negotiate allocation of the remaining 5 per cent if that target were adopted at Kyoto. As it turned out there was no need to do so, since the Kyoto meeting only adopted an overall target of a 5.2 per cent reduction, with the EU accepting a bubble target of 8 per cent.

The 1997 allocation had only covered three greenhouse gases and, since the Kyoto commitment was for six, the allocation had to be renegotiated in 1998. The UK Council Presidency consulted all states and proposed an allocation in June 1998. Council members then 'hammered out an agreement after two days of difficult negotiations' (Cass 2006, 195). Under the Burden Sharing Agreement, targets range from reductions below 1990 levels of 21 per cent by Denmark and Germany, to increases of 25 per cent and 27 per cent for Greece and Portugal (Lacosta et al 2007, 225). Greece had been allowed a 30 per cent increase in the 1997 allocation, while Portugal had been allowed 40 per cent. They attempted to keep those higher targets in the 1998 negotiations but were unsuccessful, in part because Germany insisted its target be lowered from the 1997 figure of 25 per cent to 21 per cent. Nevertheless, Germany was responsible for 85 per cent of the total EU 8 per cent reduction (Vogler 2011, 160), which undoubtedly helped the process of reaching agreement, since that was what allowed the southern state increases. Although very much influenced by the pre-existing commitments which had already been made by lead states such as the UK and Germany, Vogler (2011, 160) states the final agreement was not simply a lowest common denominator result: 'A real overall reduction was accepted as the necessary basis for climate leadership.' Using only consensual decision-making, and without financial compensation, allocation of an over-all reduction had been explicitly negotiated.

The second allocation was finalized at the 'tense' European Council meeting of 11–12 December 2008, after threatened vetoes by Italy and the eastern states, led by Poland (Vogler 2011, 167-8); the Commission news release (2008) stated: 'agreement has been reached on legally binding targets … there has been some tough negotiation.' This time, three sets of reductions had to be negotiated as part of the EU 20/20/20 package of a 20 per cent reduction in GHG emissions by 2020: emission reductions from sources not covered by the European Trading System (ETS), which had begun in 2005; an increase to 20 per cent of total energy (not just electricity) coming from renewables; and improvement in energy efficiency (conservation) by 20 per cent. Since 40 per cent of total EU emissions were by then subject to the Emissions Trading System (ETS), the targets allocated amongst Member States were for the remaining 60 per cent non-ETS emissions and for each country's share of the renewable energy increase. Again, the allocation was made necessary for the EU proposals to have credibility at an upcoming CoP (the December 2009, Copenhagen meeting) but due to the addition of eastern high-emission states (although making up a relatively low portion of the EU total, and not particularly powerful within EU decision making) the cost differential challenge was greater than it had been in 1997. EU officials stated in interviews that at the outset they saw reaching agreement with the new central and eastern European Member States as their greatest challenge (Hayden 2011).

Meeting that challenge was made considerably easier by the fact that, unlike 1997, heads of state endorsed the 20/20/20 objectives at the 9 March 2007 European Council, *before* any negotiation of allocations had even begun, largely due to pressure from the lead states of Germany, France and the UK. They had already made similar domestic commitments and were happy to see them imposed upon their EU competitors, as well as being committed to an EU leadership role at Copenhagen (Hayden 2011). At that meeting, heads of state asked the Commission to develop 'equitable and cost efficient ways of sharing the burdens and effort required to attain the 20-20-20 target' (Vogler 2011, 167). The Commission proceeded to do so, this time without the use of Triptych-style technical analysis, but with financial compensation, presenting a recommended allocation in January 2008, which had 'something in it for everybody' (EU official, cited in Hayden 2011, 12). Most significantly, less developed states were again allowed targets for increased emissions, but this time accompanied by financial compensation, to be provided by

10 per cent of ETS revenues (with another 2 per cent mainly going to Romania, Poland and Bulgaria, which allowed a veto to be avoided at the final negotiating session: Vogler 2011, 168).

The Commission used the two principles of solidarity (all must make some contribution) and capability (rich countries must do more than poor ones) as the basis for its allocation (Stephenson and Boston 2010). The non-ETS allocation was decided solely on the basis of GDP per capita, but with a cap on any increase of 20 per cent above 2005 levels. Those reductions were to be achieved by Member States, but regulation by means of the ETS itself has been transferred from Member States to the EU. The renewable energy allocation included both a requirement that all Member States increase renewables to 5.5 per cent of total energy use, and then added different targets for each Member State, again based on GDP per capita. The resulting range is from Sweden, with a target of 49 per cent share of its energy to come from renewables by 2020, to Malta with a 10 per cent share target (Hayden 2011; Stephenson and Boston 2010).

We now discuss the influence of the three sets of factors and sub-factors discussed above.

Institutions

The stark economic difficulties faced by the EU since the 2008 recession, largely flowing from the fact Eurozone members share a currency but not the institutions of a federated state, remind us that the EU is a federated system, but not a federated state. It does, however, have many of the attributes of one, including the ability to make and implement policy at the 'national government' level, both by directing Member States and itself directly regulating non-state actors located within their borders. Subnational governments (Member States) have jurisdiction over environment, but over the years that has become a case of shared jurisdiction. 'The Single European Act (SEA) [effective 1 July 1987], the Treaty on European Union (TEU) [effective 1 November 1993] and the Treaty of Amsterdam (ToA) [effective 1 May 1999] have cumulatively embedded environmental protection in the EU's policy portfolio through the use of qualified majority voting (QMV) and the increased participation of the European Parliament (EP)' (Sbragia, 2000, 294).

Thus to the extent climate change is considered an environmental issue, the EU has a mandate to act, sharing jurisdiction with Member States. However, at heart climate change is an energy issue and here the EU has less certain jurisdiction: 'Despite the wave of enthusiasm for a stronger EU role in energy policy ... only the responsibility for market liberalization – and for some science, technology and innovation aspects – currently resides at the EU level' (Hildingsson et al 2010, 103). The 2009 Lisbon Treaty made energy policy also a field of shared jurisdiction, but at the time of the allocation decisions considered here, Member States held primary jurisdiction for energy policy, increasing their ability to resist national policy.

Although not represented in the European Parliament, Member States fully participate in EU-level decision making in both the Council (ministers) and European Council (heads of state). Qualified majority voting is now used in the Environment Council, but consensus is preferred and both allocation decisions were made by consensus. As discussed below, in March 1997, Germany, the UK and Belgium used their veto powers to block the Dutch presidency proposal for allocation of a 15 per cent reduction. In 2007–8, consensus was legally required for these reasons: 1) to the extent the decision was seen as energy-related; 2) because consensus was seen as necessary for legitimacy reasons; and 3) because the final decision was moved to the European Council (heads of state) which functions only on the basis of consensus (Hayden 2011, 9).

Although, in theory, the EU could have allocated burdens by its own unilateral actions (which to some extent are intergovernmental), both its limited financial resources and political realities excluded that option from any serious consideration (Haug and Jordan 2010, 92). Thus we see that EU institutions do not support the hypothesis above for subnational government success in resisting – it had jurisdiction, was fully represented in the Council and European Council, and decisions were made by consensus. Nevertheless, resisting Member States did not block cost allocation, showing institutions alone cannot fully explain the EU case.

Role of the Subnational Government

Vogler (2011) gives a picture of three groups of Member States immediately prior to the 1997 burden-sharing negotiations: 1) the lead states of Germany, Netherlands, Austria, Denmark, Sweden and Finland, whose collective emissions formed a large part of the EU total, which had already started to take domestic action and were happy to see EU policy which would impose similar costs upon others; 2) a middle group, consisting of the UK, France, Belgium, Luxembourg and Italy which had 'mixed motives' and were unlikely to play either a lead or veto role; and 3) the less industrialized states, Greece, Ireland, Portugal and Spain, which had low per capita emissions and contributed a low portion of total EU emissions but which, like the global South, were committed to economic growth and leery of any climate policy which might constrain that goal. The latter group did not play a veto role for two reasons – their emissions formed a small portion of the total and so they could be allowed increases without undercutting over-all policy effectiveness (given the policy of explicit subnational allocation), and they lacked political power due to their small size. Nor did the industrialized, high per-capita emission states play such a role. At the end of the day, Germany, already making substantial reductions for reasons unrelated to climate policy (due to re-unification and the need to modernize the former East Germany industrial stock), agreed to absorb 85 per cent of the total EU reduction to 8 per cent below 1990 levels agreed to at Kyoto – a subnational sugar daddy which did not exist among the Lander of Germany itself, nor within Canada.

The 2007/8 process faced a potential veto by the new eastern states such as Poland, which relied upon coal for 90 per cent of its electricity generation. This was seen by EU officials as the greatest challenge to development of the 2008 climate and energy package (interviews cited in Hayden 2011). As discussed, during final negotiations in December 2008, the threat of veto by those states required provision of financial compensation, in addition to that earlier promised, which in turn was additional (as had been the case in 1997/98) to the emission *increases* allowed for the less developed states. However, they did not have significant motivation as measured by per capita emissions – Poland in 2008 had only the twelfth highest emissions, at 9.9 megatonnes per person, compared to Germany at 11.2 (European Environment Agency 2009). Nor did Poland have the financial resources needed to exert significant political power, with GDP per capita in 2008 of 9500 Euros, compared to Germany's 30,200 (Eurostat 2011). In 2008, it had the sixth largest share of total emissions – 8 per cent – compared to Germany, which had the largest share at 19.4 per cent. Other eastern Member States, such as Romania (3 per cent) Lithuania (0.5 per cent), Slovenia (0.4 per cent) were less important for overall effectiveness (European Environmental Agency 2011).

Thus we see that the southern Member States in 1997/8, and eastern Member States in 2007/8, were not highly motivated by cost differentials, lacked financial resources to exert political power and were far from contributing the bulk of total EU emissions. Unlike institutions, the subnational government role does support the hypothesis provided above, and shows the importance of this factor.

Role of the National Government

As discussed, the EU bodies playing the federal government role were different in the two allocations, but in each case those involved were strongly motivated to achieve effective policy. Schreurs and Tiberghien (2007) point to this ability of different EU Member States and governing bodies to play lead roles at different times as a factor explaining policy effectiveness. Three factors explain the high motivation of the EU 'national government.' The first is a desire for global leadership in the UNFCCC regime (Hayden 2011). Second, climate policy was seen by bodies such as the Commission as another means of furthering its basic goal of European integration. Finally, both EU governing bodies and Member States shared a desire to increase use of renewables as a means of decreasing reliance upon imported Russian natural gas. EU 'national government' motivation was certainly stronger than that of its Canadian counterpart, and perhaps also stronger than that of German governments.

In both allocations, EU governing bodies were adroit in their ability to broker agreements on cost allocation. In part this was due to such things as use of the Triptych study and, in 2007/8, to the Commission ability to consider the needs of all states while assembling the Energy Package (interviews). Beyond that, the EU use of an explicit allocation opened the door to the use of instruments for making costs more equal – both by varying targets and, in 2008, through financial compensation. With respect to both motivation and use of strategy, the EU 'national government' role supports our hypothesis.

CLIMATE CHANGE POLICY IN GERMANY

In 1990 the national government led by Herman Kohl adopted a goal of a 25 per cent reduction and introduced the electricity feed-in-tariff program that was to be a central component of renewable energy policy. In that same year, the unification with East Germany was finalized, leading to 'wall-fall' reductions, resulting from plant modernization in the former East Germany unrelated to climate policy. This gave policy makers confidence they could achieve significant reductions. In March 1995, just before COP1 in Berlin, the national government, using the threat of regulation, signed an agreement for voluntary action by 19 industrial sectors, which committed to reducing emissions intensity by 20 per cent from 1990 levels by 2005. Voluntarism, supplemented by the 1998 energy tax, remained the main policy instrument used to reduce industrial emissions until, against its wishes, Germany was forced in

2005 to initiate the EU regulatory cap-and-trade program (ETS). Since reductions were voluntary, there was no pressing need for territorial allocation. During the 1990s, Lander also developed their climate policies, with North Rhine-Westphalia, in particular, providing subsidy to renewable energy (one third of the total subsidies provided by Lander) by 1998 (Monstadt and Scheiner).

A major factor with respect to the allocation challenge in German climate policy has been the fact that emissions in the former East Germany have declined significantly due to national government capital spending on plant modernization. To a certain extent, Germany, like the UK after the switch from coal to natural gas, has been given a free ride in the area of climate policy. Analysts generally attribute roughly a little under half of German emissions reduction to the exogenous factor of unification – Michaelowa (2008) says 30 per cent of reductions can be attributed to unification, while Hatch (2007) gives a figure of 47 per cent. Due to the economic crash in the former East Germany, total German emissions in 1998 were already 14 per cent below 1990 levels (Ziesing 2009, 346). The likelihood of eastern high energy-intensity Lander resisting national policy was reduced for this reason.

The Social Democrat-Green government led by Chancellor Gerhard Schroeder took power in 1998 with a commitment to increased effort on climate change. This was expressed in the October 2000 First National Climate Change Mitigation Program, which included a new voluntary agreement between the national government and the Federation of German Industry and individual associations, and which listed reduction targets for seven sectors: households; transport; industry; energy; renewables; waste management; and agriculture (Weidner and Eberlien 2009, 319). Sectoral allocation was explicit, but territorial allocation was not. In addition, a new energy tax was imposed on electricity consumption (with reduced rates for industry and exemptions for coal and nuclear fuels), and a revised feed-in-tariff program was introduced by means of the Renewable Energy Act (REA), which came into effect 1 April 2000.

The Renewable Energy Act was the first national government policy to address the issue of cost allocation amongst the Lander. In previous years, wind-power installation had been increasing faster in some regions, particularly those in the northern coastal areas, than others, imposing additional subsidy costs on those utilities. In response to northern Lander complaints, the REA required pooling and sharing of those costs amongst utilities in all the Lander. Thus, while following a policy of implicit geographic allocation, the German government did provide financial compensation.

When the EU released its first Green Paper on emissions trading in 2000, the German national government expressed its opposition, because it would be compulsory, while Germany was relying on voluntary agreements with industry, and because it would impose higher costs on German industry than those associated with the previously negotiated voluntary agreements: 'Germany was actually one of the countries that blocked or at least delayed the introduction of the [EU] ET directive for a long time' (Ziesing 2009, 349). The Lander, particularly those with high per capita emissions such as North Rhine-Westphalia, supported the national government in that opposition. Unable to block the ETS completely, the German national government eventually agreed to it, after negotiating a compromise that would allow opting out for specific sources (Ziesing 2009, 349). The EU Directive requiring Member States to develop national trading systems was then issued in 2003. As it worked to develop its trading plan, the German national government was heavily lobbied by the industrial sectors and also faced demands by the Lander that they, rather than the national government, implement the program, since that would allow them to better protect their emission sources. North Rhine-Westphalia and other large Lander tried to gain that control in the Bundesrat as the national law was being passed by both houses, but were outvoted by smaller Lander. Interviews reveal that those Lander failed to support NRW both because their lower energy intensity meant the ETS was less of a threat, and because they lacked the administrative capacity for the ETS policy file (Monstadt and Scheiner).

As a result of industry and utility lobbying, Germany issued too many trading permits in the first round, making the ETS largely ineffective. 'As result of a highly political process with heavy utility and industry lobbying, German implementation produced a bureaucratic system of emissions permit allocation that [was] heavily biased in favour of specific utility and other industry interests and required only marginal emission reduction rates' (Zeising 2009, 349). The system was improved in the second round, 2008–12, and Germany strongly supported the new ETS introduced as part of the 2008 20/20/20 package. Its one area of opposition was to increased motor vehicle efficiency requirements. Supported by those Lander in which the auto industry is located (Bavaria, Baden-Wurttemberg, Hessen, North Rhine-Westphalia and Lower Saxony), Germany forced a weakening of that EU requirement.

Institutions

Germany is a federated state in which the national government is composed of two bodies, one, the Bundestag, representing the people, and the other, the Bundesrat, representing the 16 subnational governments, the Lander. The Chancellor, who is elected by the Bundestag, forms the national government, often through party coalitions, since Germany has a proportional representation electoral system. National government policy that affects the Lander, which includes most environmental and climate change law, must be approved by both the Bundestag and the Bundesrat (Schmidt 2003). The other major institutional feature of German federalism is the fact of membership of the European Union, which means that a significant part of German climate change policy, particularly in more recent years, is decided at EU level, rather than by German national or subnational governments. As set out above, Germany has always played a substantial role in EU climate change decision making. The Lander have a constitutional right to participate directly in EU affairs, but have little influence there (Monstadt and Scheiner).

The Lander have jurisdiction, in terms of a mandate to implement policy within their own borders, over energy and climate change policy. Particularly in the 1980s, it was the Lander rather than the national government which took the lead in developing renewable energy (Monstadt and Scheiner). Beginning in the mid-1990s, and continuing today, almost all Lander have their own climate policies, which are implemented independently of national government policy (Monstadt and Scheiner). Nevertheless, the main components of German climate policy – the feed-in-tariff, measures in successive national plans, and the trading system – have come from national government policy. That policy is for the most part implemented directly by the national government, unlike most other policy areas in which Lander implement national government policy (Weidner and Mez 2008, 371). This too has the effect of reducing subnational government ability to resist, through delays in implementing national policy: 'the 16 federal Lander have almost no implementation responsibilities with respect to central climate change policies. Therefore, implementation deficits caused by specific Lander interests and capacities generally do not arise' (Weidner and Mez 2008, 371).

Lander participate in national government climate policy making through their roles in the Bundesrat, which makes decisions by majority voting. The Lander are given votes in the Bundesrat in proportion to their size, with six votes each for the largest, four for the mid-range and three each for the smallest. Voting there is influenced both by territorial interests and by party interests. If the opposition party, as a result of

Lander level elections, which are often seen as a vote of confidence in the national government, holds a majority in the Bundesrat it will use that power to influence national government policy (Silvia 1999).

In a limited sense, the German federal system also provides for intergovernmental relations outside the domains of the national and Lander governments. Established in 1973, the Conference of the Environmental Ministers, made up of the federal and Lander ministers and supported by a small, two-person secretariat, meets twice a year. The primary purpose is to allow exchange of information. Resolutions are adopted by majority vote but are not binding. 'Relative close and continuous cooperation [on climate policy] between central and Lander governments are aided by a biannual Conference of Environment Ministers and central-state government working groups' (Weidner and Mez 2008, 371-2). The intergovernmental system is not, however, the forum in which climate policy decisions are made.

Thus we see that the Lander have climate policy jurisdiction but the more significant policy measures are implemented by the national government. The Lander directly participate in national government decision making, but it is difficult for the potentially resisting large northern Lander to achieve a majority in the Bundesrat. Finally, decision making is by majority vote, rather than consensus. German institutions support the hypothesis provided above.

Role of the Subnational Government

Within Germany, GHG emissions are concentrated primarily in the north, home to older industries of coal, steel, textiles. The Lander of North Rhine-Westphalia (NRW), Saarland and Brandenburg, in which German coal-mining is concentrated, have worked to influence federal policy makers as they have negotiated the allocation of costs with industrial sectors, both through voluntary programs and the design of an emissions-trading system (Huber 1997; Lauber and Mez 2004; Cass 2006). Per capita emissions in NRW are 16 tonnes, compared to the Lander at the other end of scale – Thuringia, 4.8 tonnes; Berlin, 5.1; Schleswig-Holstein, 6.6 and Bavaria, 6.4 (Macdonald et al 2013). This ratio of approximately four to one perhaps makes NRW less motivated to resist than does the Alberta ratio of seven to one, discussed below. NRW was responsible for 34.7 per cent of total German emissions in 2008, meaning it could not be ignored (NID 2012). In that same year, while not the highest, it had GDP per capita of 30,300 Euros, compared to the lowest Lander, Mecklenburg-Vorpommern which had 21,200 (Eurostat 2011).

This suggests it has the economic capacity to wield influence in German decision making.

Despite having at least some incentive and the means to resist national policy, beyond the one instance recounted above, it does not seem that NRW has consistently played that role. Weidner and Metz (2008, 372) give this assessment: 'Although the degree of support for climate policy varies from state (*Land*) to state (*Land*), there is no state that stands out for its opposition to federal initiatives.' This lack of subnational resistance, despite at least some motivation and, setting aside the institutional context, political power in the federated state tends not to support the hypothesis above.

Role of the National Government

In terms of degree of motivation, we find a consistent commitment from successive governments, with the ideology of the governing parties having little influence. Climate policy of the 1999–2005 government of Social Democrats and Greens has been largely continued by the CDU-SPD government of Chancellor Angela Merkel. While consistent, that motivation has not been as high as that of the EU governing bodies. It has always been heavily influenced by industry lobbying, and for that reason has preferred voluntary measures (with the important exception of the feed-in-tariff). The combination of the 'wall-fall' reductions achieved through unification, pressure from the EU, a strong environmental movement grounded in supportive public opinion, and an ecological modernization discourse that saw economic benefit in climate action, as well as cost, has meant that successive national governments have remained committed to effective policy. It seems fair to classify German government motivation as 'mixed' – not as high as the EU, but much higher than that of the Government of Canada.

As discussed, the national government adopted a strategy of implicit rather than explicit allocation of cost amongst subnational governments.

> Both the government and the proponents of a strong global climate change policy have provided the public with considerable information about the net positive benefits for the country as a whole of its climate change programmes, but only with sparse information about the expected distributional effects (costs and benefits) of present and planned policies and the international commitments for specific groups and regions. In sharp contrast to the salient role global equity concerns are playing in the political discourse on climate change, the issue of equity within Germany has been almost completely neglected by most proponents of progressive climate policy as well as academic circles (Weidner and Metz 2008, 374).

Germany had the luxury of such a strategy, which presumably the national government of any federated system would prefer, since it avoids negative electoral repercussions. The EU, as a system of sovereign states, did not have that luxury. Canada, as we see below, mistakenly thought it too could use such a strategy but in its case it has been a recipe for inability to achieve any form of national allocation, sectoral or geographic. With sufficient and consistent motivation, and using a strategy suitable for the context (which in the case of the REA included financial compensation) the national government role supports our hypothesis.

CLIMATE CHANGE POLICY IN CANADA

As described in more detail in Chapter 8, during the ten-year period following Canadian ratification of the UNFCCC in 1992 and ending with ratification of the Kyoto Protocol in 2002, an attempt was made to develop coordinated national policy, using the vehicle of periodic joint meetings of federal and provincial environment and energy ministers, supported by a complex network of committees staffed by federal and provincial civil servants. That process generated a series of federal-provincial agreements, most notably the 1995 National Action Program for Climate Change and the 2000 National Implementation Strategy and Business Plan. Due to Alberta's pressure, but also because of a lack of appetite for effective action among all governments, the policy instrument of voluntary action was used, rather than more effective regulatory or economic instruments. Although the criterion of equity for cost allocation was adopted by First Ministers when in 1997 they agreed no region should bear an 'undue' portion of the total cost, no serious effort was made to explicitly address the issue of cost sharing amongst provinces, despite the fact that all concerned knew this was the most important issue threatening success of the intergovernmental effort (confidential interviews, Macdonald et al 2013). This was because all involved, with the exception of Quebec officials, believed explicit allocation amongst provinces was too divisive and that agreement should be reached first on easier issues.

Ratification of the Kyoto Protocol in 2002 by the national government precipitated Alberta's withdrawal from the intergovernmental relations process in the spring of that year. In effect, though, the national government had already withdrawn by then, since it was contemplating unilateral regulation regardless of provincial action. In the fall of 2002, the provincial premiers requested a meeting of First Ministers (Prime Minister and Premiers) prior to Kyoto ratification. The national government refused the request, ratified Kyoto, and national and provincial ministers ceased to meet. From 2003 to 2006, the national government for the first time developed its own unilateral climate change policy, working directly to reduce emissions from sources through the instruments of subsidy, and negotiating with the largest firms for reductions that would be required by federal law.

Since assuming power in January 2006, the national government led by Stephen Harper has continued the plans of the previous government to directly regulate industrial sources, but has abandoned any effort to bring about coordinated national policy (Macdonald 2011). Instead it is coordinating its own policy with (or more accurately mimicking) US federal government climate policy while leaving the provinces to do what they wish. In 2010, following the Copenhagen Conference of Parties, the national government adopted the same target as the US – reduction of GHG emissions to a point 17 per cent below the 2005 level by 2020. A number of provinces, acting in isolation or coordinating action with US states through such things as the Western Climate Initiative, have begun to put in place more effective policy measures, such as a carbon tax of $30 tonnes in British Columbia and a feed-in-tariff program for renewable energy in Ontario. It is estimated that during the period leading up to 2020, the current and planned reduction policies of all governments will only achieve half of the 17 per cent reduction target (NRTEE 2012, 112). An independent policy think-tank, the National Round Table on Environment and Economy report (2012, 115) states that the present completely uncoordinated approach is doomed to failure:

> No one sector and no one province can make up all the difference [the missing half of reductions needed to achieve the target]. This puts a premium on intergovernmental collaboration and coordination of measures.

Why has Canada, unlike the EU and Germany, been unable to agree on allocation of GHG reductions among both sources and subnational jurisdictions? We now examine the ways in which our three factors help to answer that question.

Institutions

As set out in the original British North America Act adopted by the UK Parliament in 1867, Canada is a federated state, with enumerated powers for each of the two levels of government and residual powers granted to the federal government. Due to a series of British court decisions in the nineteenth century which favoured the provinces, and, since the 1960s, the very real threat of Quebec separatism, it is today one of the world's most decentralized federations, perhaps exceeded only by Belgium. Canadian provinces, in particular those with sufficient population and financial capacity to mount policy and legal challenges to resist federal government intrusion (British Columbia, Alberta, Ontario and Quebec) are largely autonomous in the fields in which they have jurisdiction, such as health and education. Unlike the EU and Germany, which have the recognized power to require policy action by Member States or Lander, the Government of Canada, in *de facto* terms, cannot directly require provinces to act, and instead can only attempt to induce such action by offers of new funding or in some cases (effectively limited to health policy) by threats to withdraw existing funding.

Provinces clearly have jurisdiction in the realm of the environment and energy which constitutes climate change. Because the Canadian constitution gives them ownership of resources, it was initially assumed the provinces had primary jurisdiction over environment and federal policy action was resisted by the large provinces. Supreme Court decisions have since decreed that both levels of government have a constitutional mandate to make policy in the field of environment and, although this has not been tested in the courts, it is assumed this is also true of climate change (Chalifour 2010). However, other than at times when popular concern for environment is spiking in the polls, offering electoral rewards for the appearance of action, the federal government largely defers to the provinces and leaves the policy field to them (Harrison 1996; Weibust 2009). In 1991, provincial and local government spending accounted for 90 per cent of total environmental policy spending (Skogstad 1996) and there is no reason to believe that would be different today. Three-quarters of current climate change policy is being implemented by the provinces, with only the remaining quarter by the national government (NRTEE 2012).

In terms of the second aspect of institutions examined here, Canadian provinces have no ability to participate directly in national government policy processes. This is because the upper house, the Canadian Senate, does not provide regional representation and, as an unelected body, while it reviews legislation enacted in the House of Commons, it has no real legislative powers. This means co-ordinated policy can only be developed through the process of intergovernmental relations. Since provinces are under no legal obligation to participate, nor subject to any strong political pressure to do so, each can threaten to opt out of the process at very little cost. This means decisions can only be made by consensus, driving policy effectiveness down to the lowest common denominator.

In summary, the institutions of Canadian federalism give subnational governments primary jurisdiction respecting climate change. They cannot directly participate in national government decision-making but pan-federation policy is not made there; and intergovernmental decision making, the forum available for pan-Canadian policy making is done only by consensus. Clearly institutions help explain Canadian inability to allocate GHG reductions and support the hypothesis above.

Role of the Subnational Government

The ratio between highest and lowest per capita emissions amongst Canadian provinces in 2009 was roughly seven to one: Saskatchewan had emissions of 71 tonnes per person, Alberta 63.7; while at the other end of the scale, Ontario emissions were 12.6 tonnes per person and those of Quebec 10.4 (Environment Canada 2011). This is largely due to the fact that Alberta and Saskatchewan both produce oil and natural gas, an energy-intensive activity, and also rely upon coal for a considerable portion of their electricity generation. While both Ontario and Quebec are highly industrialized, with associated emissions, both have access to hydro-electricity, supplemented by emissions-free nuclear power. From the beginning, it has been recognized that national climate change policy will impose significantly higher costs upon Alberta and Saskatchewan than other provinces, and not surprising they have been the most resistant.

Per capita Alberta GDP in 2008 was $72,110 (Statistics Canada). That in Saskatchewan was $60,072, compared to those at other end of scale; Nova Scotia $36,892 and PEI $33,541 (Statistics Canada). The two resisting subnational governments clearly hold substantial financial resources.

Alberta accounts for 34 per cent of total Canadian emissions and Saskatchewan another 11 per cent, meaning that together they are responsible for just under half of Canadian emissions and so must play a significant role if meaningful reductions are to be made in Canada. Those two provinces have been the most resistant to national policy action, playing a veto role to stall or block policy until 2002 and subsequently introducing ineffective policies. However, Alberta was the first Canadian government to use law, rather than voluntarism, to regulate industrial emissions (enacted in 2002, largely to prevent the federal government from using law to regulate Alberta sources). But standards are low and the bulk of Alberta emissions reductions are expected to come from carbon capture and storage (CCS). Since the effective price of carbon in Alberta is $15 a tonne and the cost of CCS is perhaps ten times that, the likelihood of significant Alberta reductions in the near term is low. Saskatchewan has not made any effort to adopt policy instruments such as law, tax or feed-in subsidy. An environmental NGO (Suzuki Foundation 2012, 2–3) ranks Ontario, British Columbia and Quebec as the provinces with the most effective climate change policies, and says: 'With the country's highest emissions, and relatively weak laws and incentives for clean energy, Saskatchewan and Alberta remain at the bottom [of the ranking].'

In Canada, the resisting provinces are motivated by reduction cost differentials, have political power and, given their share of total emissions, cannot be ignored. In this instance, the hypothesis is supported.

Role of the National Government

The year Canada signed the UNFCCC, 1992, marked one of the high points of popular support for environmental action and the corresponding federal government movement into the policy field, seeking electoral advantage. Not surprisingly, the national government took the lead in coordinating the federal-provincial process that led to the 1995 national program. By 2002, however, it had lost faith in a process which it thought was unlikely to ever rise above the lowest common denominator of voluntarism (interviews) and opted instead for unilateral regulation. There was a significant lessening of federal government motivation in 2006, when the Harper government, ideologically anti-environmental and with an electoral base in western Canada, took power. All national governments have followed a policy (with the sole exception of Kyoto ratification) of ensuring their policy does not differ markedly from that of the US government –which given the lack of action there has reduced national government motivation.

The Liberal government from 1992–2002 tried to broker a national program on the basis of implicit allocation, because, as noted, officials believed an explicit approach would be divisive, threatening national unity and possibly increasing barriers to inter-provincial trade (interviews, Macdonald et al 2013). Although officials were aware of the need to consider financial compensation, such as national government funding of job retraining for workers affected by climate policy, no real effort, driven from the top ministerial level, was ever made to develop such programs (interviews, Macdonald et al 2013). During that period, had it been more motivated, it could have done more to induce provincial action by threatening unilateral regulation (using that threat to gain bargaining leverage, rather than simply adopting that policy in 2002). It could also have stayed with the federal-provincial process rather than effectively ending it with Kyoto ratification in 2002 (it appears Prime Minister Chretien was more interested in achieving the symbolic action of ratification before he left office than in using the threat of ratification to move the federal-provincial process forward). More importantly, it failed to recognize that subnational jurisdiction allocation already *was* explicit, since it was at the top of the mind always of those developing coordinated policy in the 1990–2002 period.

A national government with weak motivation, using an inept strategy, confirms the hypothesis set out above.

CONCLUSION

As set out in the introduction, it is hoped that a comparison of these three cases will tell us something about the ways in which the three factors examined contribute to success in allocating GHG reduction costs in federated systems. We now discuss each of the three in turn.

In terms of institutions, Germany is distinct from the other two cases, since only in that case was policy made by the national government through use of majority voting, rather than in an intergovernmental forum by means of consensual decision-making. The 2003 failure of NRW to win a Bundesrat vote respecting Lander implementation of the ETS shows how the institutional context of majority vote and weighted votes in Germany stacked the deck against subnational governments seeking to influence national climate change policy. Differences in the institutions of federalism, rather than roles of subnational or national governments, is most important for explaining allocation success in Germany compared to failure in Canada. Institutions cannot, however explain EU success compared to Canada, since both relied upon consensual intergovernmental decision making. Given the similarity of EU and Canadian institutions, we must conclude (with a caveat expressed immediately below) that in this comparison institutions do little to explain the different outcomes. Since institutions explain one comparison but not the other, we must conclude they cannot by themselves fully explain the ability of federated systems to allocate GHG reduction costs.

The caveat arises from the fact, discussed above, that the institutional context in the EU varied from the first to the second allocation. The EU Commission had attempted to lead the allocation process in 1996 and had failed, with the result the allocations of 1997 and 1998 were done through an intergovernmental process, albeit with leadership provided by the Presidency. As noted, that allocation involved difficult negotiations, but did not impose costs significantly greater than had been previously accepted by Member States. It is very possible an attempt to allocate the full 15 per cent reduction would have led to Canadian-style failure. The 2008 allocation, which did more to impose costs which had not been previously accepted by states and which faced a greater cost differential challenge, was much more a top-down process (Vogler 2011, 152). Heads of state at the March 2007 meeting endorsed the targets before allocation was considered, and the Commission was then able to play an active role, negotiating a package, complete with compensation as needed, acceptable to all. Haug and Jordan (2010) argue this was in part due to the overall increase in EU level authority which had occurred during the intervening ten years. This difference supports the finding from the German case, that institutions of centralized governance make allocation more likely – supporting the Weibust (2009) argument that centralization is related to policy effectiveness. While this caveat is important, the fact remains that similar institutions produced two different results and so we must look elsewhere for a full explanation.

We now turn to the role of the subnational government. Here, the Canadian case clearly shows the importance of this factor. Alberta and Saskatchewan are highly motivated, stemming from the greater cost they would be asked to bear, relative to other provinces. Since they were responsible for a large portion of total emissions during the attempt to develop national co-ordinated policy, 1990–2002, they could not be ignored by other governments without unduly sacrificing policy effectiveness (which is exactly what has happened since). And third, Alberta, in particular, due to its economic strength holds substantial political power within the federated system. These powerful subnational governments first delayed agreement on national policy, including cost allocation, and then Alberta, when it left the process, guaranteed it could not be put into place (although, as discussed, Alberta formally left in 2002 because the federal government had by that time informally abandoned the process). The 2008 EU case also confirms the importance of this factor, since the eastern Member States with lower motivation, lower shares of total emissions and much less power in the federated system were unable to stall the process in the way Alberta and Saskatchewan did. The German case, on the other hand, does not give such a clear reading. NRW had a lower cost differential than Alberta with accompanying lower motivation, but it had a comparable share of total emissions and economic strength. Given that it had comparable political power, why could it not exert the same influence as Alberta? This points to the importance of subnational government degree of motivation as an explanatory factor – perhaps it simply did not try as hard. However, it also underlines, again, the importance of institutions. No matter how motivated, the fact that it had only its votes in the Bundesrat, instead of the option of withdrawing from a consensual IGR process, meant it never had a chance of completely stalling the allocation process the way its Alberta counterpart did.

Third, we consider the role of the national government. The importance of this factor is shown by the comparison of EU success with Canadian failure. The EU governing bodies were highly motivated and used adroit strategies to broker agreement on both occasions, using both differing targets and financial compensation to make costs more equal. During the period that it was actively attempting to develop national policy, 1992 to 2002, the Government of Canada, by comparison, was much less motivated. It had no aspirations to global leadership (quite the reverse, it wanted only to be sure Canadian policy did not differ too much from American and saw global leadership as a threat to internal unity rather than means of achieving it). It refused to consider financial compensation as a means of equalizing costs and then, the most maladroit strategy of all, abandoned the IGR process in 2002, deciding it would regulate unilaterally. Given the similarity of institutions, this difference in motivation and strategy of the national government goes a long way towards explaining EU success and Canadian failure. Hayden (2011, 17) puts it this way: 'When asked about the lessons they saw for Canada from the EU's ability to agree on emissions-reduction efforts, one respondent said, "the main thing is to have that political ambition in the first place."'

The national government role does less, however, to explain German success compared to Canadian failure. As we have seen the German national government had at best mixed motivation – its task was made easier by the unification 'wall-fall' reductions; it preferred voluntary measures over regulation and resisted and weakened the EU trading system. Nor did it need to use strategies in an IGR forum, since German policy was not made there. The German government succeeded despite relatively weak motivation because of the power given to it by the institutional context – national government decision making by majority voting.

Where does that leave us? With exceptions (EU institutions and the German role of subnational governments), the hypothesis explored here has been confirmed. Not surprisingly, we have found that institutions matter. They almost completely explain one comparison (Germany v Canada) and partially explain another (EU 2008 v Canada). They also explain two aspects of the German case, since institutions gave success, despite the importance and power of a resisting subnational government (one exception to the hypothesis) and a national government with mixed motivation. This finding confirms the focus in the federalism literature, and more specifically the finding by Weibust (2009) that, despite claims to the contrary, subnational governments by themselves cannot achieve effective co-ordination in the area of environmental policy.

However, the EU achieved success despite the assumption in the hypothesis that its institutions would generate failure. This confirms that institutions do not provide a full explanation. For that we must also look to the motivation, powers and strategies of the relevant subnational and national governments. To some extent, this is good news for federated systems, which in future must face the inherent challenge of allocating reductions amongst sources and subnational jurisdictions. Those governments have relatively little ability to change their federal institutions in the short term. However, they *can* adopt effective strategies – in particular, when necessary, to adopt explicit allocation with use of side-payments to make costs more equal.

ACKNOWLEDGEMENT

This paper is the product of a research effort funded by the Social Sciences and Humanities Research Council of Canada: 'Allocating Canadian greenhouse gas emission reductions amongst sources and provinces: learning from Germany and the EU' 2009–12, Douglas Macdonald, University of Toronto; Jochen Monstadt, Technische Universitat Darmstadt, Kristine Kern, University of Potsdam

NOTES

1. The term 'federated system' is used here to refer to both a federated state and the EU. To simplify discussion, although technically incorrect, the term 'national government' is subsequently used to refer to the governance system of the EU (essentially Council, Parliament and Commission) and also to the Government of Germany and the Government of Canada. The problem of terminology is compounded by the fact that within the EU, the Council is a more intergovernmental body, while the Commission and Parliament are more 'national.' The Government of Canada is usually referred to as the 'federal' government but for consistency that practice is not followed here.
2. Again, while technically incorrect, for ease of discussion EU Member States are referred to as 'subnational governments.'

REFERENCES

Bakvis, Herman and Grace Skogstad (2012), 'Canadian Federalism: Performance, Effectiveness and Legitimacy,' in Herman Bakvis and Grace Skogstad (eds), *Canadian Federalism: Performance, Effectiveness and Legitimacy*, Don Mills: Oxford University Press Canada, pp. 2–19.

Bolleyer, Nicole (2009), *Intergovernmental Cooperation: Rational Choices in Federal Systems and Beyond,* Oxford: Oxford University Press.

Cass, Loren R. (2006), *The Failures of American and European Climate Policy*, Albany: State University of New York Press.

Chalifour, Nathalie J. (2010), 'Chapter 10. The Constitutional Authority to Levy Carbon Taxes,' in Thomas J. Courchene and John R. Allan (eds), *Canada: The State of the Federation 2009: Carbon Pricing and Environmental Federalism*, Montreal and Kingston: McGill-Queen's University Press, pp. 177–95.

David Suzuki Foundation (2012), *All Over the Map: A Comparison of Provincial Climate Change Plans,* The David Suzuki Foundation, Vancouver.

Environment Canada (2011), *National Inventory Report 1990 – 2009. Greenhouse Gas Sources and Sinks in Canada.*

European Environmental Agency (2011), 'Annual European Union Greenhouse Gas Inventory 1990-2009 and Inventory Report 2011, *Technical Report* No. 2/2011, Submission to the UNFCCC Secretariat, 27 May 2011.

European Environment Agency (2009), 'GDP vs. GHG per Capita per Country from 1990-2010' (Eurostat), http://www.eea.europa.eu/data-and-maps/daviz/sds/gdp-vs.-ghg-per-capita-3/@@view. Data as of 1 January 2009.

EurostatNewsrelease, 'Regional GDP in the European Union, 2008'. Reference: STAT/11/28. Date: 24 February 2011. Available at: http://europa.eu/rapid/pressReleasesAction.do?reference=STAT/11/28&type=HTML, Accessed 5 September 2012.

EU Commission, news release 17 December 2008.

Galbraith, John Kenneth (1982), *The Anatomy of Power*, Boston: Houghton Mifflen.

Harrison, Katherine (1996), *Passing the Buck: Federalism and Canadian Environmental Policy*, Vancouver: UBC Press.

Hatch, Michael T. (2007), 'The Politics of Climate Change in Germany: Domestic Sources of Environmental Foreign Policy', in Paul G. Harris (ed.), *Europe and Global Climate Change: Politics, Foreign Policy and Regional Cooperation,* Cheltenham, UK: Edward Elgar, pp. 41–62.

Haug, Constanze and Andrew Jordan (2010), 'Sharing Burdens or Distributing Efforts? Negotiating Emission Reduction Commitments in the European Union', in Andrew Jordan, Dave Huitema, Harro vanAsselt, Tim Rayner and Frans Berkhout (eds), *Climate Change Policy in the European Union: Confronting the Dilemmas of Adaptation and Mitigation?* Cambridge: Cambridge University Press, pp. 83–102.

Hayden, Anders (2011), 'Europe's Climate and Energy Policy: Lessons for Canada in Sharing the Effort of Emissions Reductions?' Unpublished paper in the possession of the author.

Hildingsson, Roger, Johannes Stripple and Andrew Jordan (2010), 'Renewable Energies: A Continuing Balancing Act?' In Andrew Jordan, Dave Huitema, Harro vanAsselt, Tim Rayner and Frans Berkhout (eds), *Climate Change Policy in the European Union: Confronting the Dilemmas of Adaptation and Mitigation?* Cambridge: Cambridge University Press, pp. 103–24.

Hoffmann, Matthew J. (2011), *Climate Governance at the Crossroads: Experimenting with a Global Response*, Oxford: Oxford University Press.

Huber, Michael (1997), 'Leadership and Unification: Climate Change Policies in Germany,' in Ute Collier and Ragnar E. Löfstedt, (eds), *Cases in Climate Change Policy: Political Reality In the European Union,* London: Earthscan, pp. 65–86.

Hueglin, Thomas O. and Alan Fenna (2006), *Comparative Federalism: A Systematic Inquiry*, Peterborough: Broadview Press.

Johansson-Stenman, Olof and James Konow (2010), 'Fair Air: Distributive Justice and Environmental Economics', *Environment and Resource Economics* **46**, 147–66.

Jordan, Andew and Tim Rayner (2010), 'The Evolution of Climate Policy in the European Union: An Historical Overview,' in Andrew Jordan, Dave Huitema, Harro vanAsselt, Tim Rayner and Frans Berkhout (eds) *Climate Change Policy in the European Union: Confronting the Dilemmas of Adaptation and Mitigation?* Cambridge: Cambridge University Press, pp. 52–80.

Kelemen, Daniel R. (2004), *The Rules of Federalism: Institutions and Regulatory Politics in the EU and Beyond,* Cambridge: Harvard University Press.

Lacasta, Nuno S., Suraje Dessai, Eva Kracht and Katharine Vincent, (2007), 'Articulating a Consensus: The EU's Position on Climate Change,' in Paul G. Harris (ed.), *Europe and Global Climate Change: Politics, Foreign Policy and Regional Cooperation,* Cheltenham: Edward Elgar, pp. 211–31.

Macdonald, Douglas, Debora VanNijnatten and Andrew Bjorn (2004), 'Implementing Kyoto: When Spending is Not Enough,' in G. Bruce Doern (ed.), *How Ottawa Spends: 2004-05*, Montreal and Kingston: McGill-Queen's University Press, pp. 175–97.

Macdonald, Douglas, David Gordon, Asya Bidordinova, Anders Hayden, Jochen Monstadt, Kristine Kern, Stefan Scheiner, Alexey Pristupa (2013), *Allocating Canadian Greenhouse Gas Emission Reductions Amongst Sources and Provinces: Learning from the EU, Australia and Germany,* report of a project funded by SSHRC, Toronto.

Macdonald, Douglas (2011), 'Harper Energy and Climate Policy: Failing to Address the Key Challenges,' in G. Bruce Doern and Christopher Stoney (eds), *How Ottawa Spends*, Montreal and Kingston: McGill-Queen's University Press, pp. 127–43.

Michaelowa, Axel (2008), 'German Climate Policy: Between Global Leadership and Muddling Through,' in Hugh Compston and Ian Bailey (eds), *Turning Down the Heat: The Politics of Climate Policy in Affluent Countries,* New York: Palgrave Macmillan, pp. 144–63.

Monstadt, Jochen and Stefan Scheiner, 'Allocating Greenhouse Gas Emissions in the German Federal System: Regional Interests and Federal Climate Governance', unpublished paper in possession of the author.

National Round Table on the Environment and Economy (2012), *Reality Check: The State of Climate Progress in Canada,* The National Round Table on the Environment and Economy, Ottawa.

Nationaler Inventarbericht Deutschland (2012), 'Germany Submission to UNFCCC,' 2012: 144. Available at: http://unfccc.int/national_reports/annex_i_ghg_inventories/national_inventories_submissions/items/6598.php

Sbragia, Alberta M. (2000), 'Environmental Policy,' in Helen Wallace and William Wallace (eds), *Policy-Making in the European Union*, Oxford: Oxford University Press. pp. 293–316.

Schmidt, Manfred G. (2003), *Political Institutions in the Federal Republic of Germany*, Oxford: Oxford University Press.

Schreurs, Miranda A. and Yves Tiberghien (2007), 'Multi-Level Reinforcement: Explaining European Union Leadership in Climate Change Mitigation,' *Global Environmental Politics*, **7** (4), 19–46.

Silvia, Stephen J. (1999), 'Reform Gridlock and the Role of the Bundesrat in German Politics,' *West European Politics*, **22** (2), 167–85.

Simeon, Richard and Amy Nugent (2008), 'Parliamentary Canada and Intergovernmental Canada: Exploring the Tensions,' in Herman Bakvis and Grace Skogstad (eds) *Canadian Federalism: Performance, Effectiveness and Legitimacy,* Don Mills: Oxford University Press Canada, pp. 89–111.

Skogstad, Grace (1996), 'Intergovernmental Relations and the Politics of Environmental Protection in Canada,' in Kenneth M. Holland, F.L. Morton and Brian Galligan (eds), *Federalism and the Environment*, Westport, CT: Greenwood Press, pp. 103–34.

Statistics Canada (2011), 'Table A.34 Gross Domestic Product per Capita, Canada, provinces and territories, 2005/2006 to 2009/2010 (in current dollars)'. Available at: http://www.statcan.gc.ca/pub/81-595-m/2011095/tbl/tbla.34-eng.htm. Accessed 5 September 2012.

Stephenson, Paule and Jonathan Boston (2010), 'Climate change, equity and the relevance of European 'effort-sharing' for global mitigation efforts', *Climate Policy* **10,** 3–16.

Vogler, John (2011), 'EU Policy on Global Climate Change: The Negotiation of Burden Sharing', in Daniel C. Thomas (ed.), *Making EU Foreign Policy: National Preferences, European Norms and Common Policies,* New York: Palgrave Macmillan, pp. 150–73.

Wallace, Helen (2000), 'The Institutional Setting', in Helen Wallace and William Wallace (eds), *Policy-Making in the European Union*, Oxford: Oxford University Press, pp. 3–37.

Weibust, Inger (2003), 'Implementing the Kyoto Protocol: Will Canada Make It?' In David Carment, Fen Osler Hampson and Norman Hillmer (eds) *Canada Among Nations 2003: Coping with the American Colossus*, Don Mills: Oxford University Press Canada, pp. 287–311.

Weibust, Inger (2009), *Green Leviathan: The Case for a Federal Role in Environmental Policy*, Farnham, UK: Ashgate Publishing.

Weidner, Helmut and Lutz Mez (2008), 'German Climate Change Policy: A Success Story With Some Flaws,' *The Journal of Environment and Development*, **17** (4), 356–78

Weidner, Helmut and Burkard Eberlein (2009), 'Still Walking the Talk? German Climate Change Policy and Performance,' in Burkard Eberlein and G. Bruce Doern (eds) *Governing the Energy Challenge: Canada and Germany in a Multi-Level Regional and Global Context*, Toronto: University of Toronto Press, pp. 314–43.

Winfield Mark and Douglas Macdonald (2008), 'The Harmonization Accord and Climate Change Policy: Two Case Studies in Federal-Provincial Environmental Policy', in Herman Bakvis and Grace Skogstad (eds), *Canadian Federalism*, 2nd ed, Don Mills: Oxford University Press Canada, pp. 266–88.

Ziesing, Hans-Joachim (2009), 'EU Emissions Trading and National Allocations Plans 2005-07: The Case of Germany', in Burkard Eberlein and G. Bruce Doern (eds), *Governing the Energy Challenge: Canada and Germany in a Multi-Level Regional and Global Context*, Toronto: University of Toronto Press, pp. 344–72.

PART IV

Findings on effectiveness and governance patterns

10. Ensuring the effectiveness of European Union environmental law: from supranational law making to multilevel enforcement

Marc Pallemaerts

Since the early 1970s, the European Union's (EU) supranational legislative machinery has developed an impressive array of environmental legislation that prevails over the national laws of its Member States. Yet it is well known that the effectiveness and enforcement of this legislation continues to pose major problems in all Member States as well as, paradoxically, at the level of the EU institutions themselves. Serious concerns about the effectiveness of Community environmental law have been raised both by academic writers and policy makers, especially since the mid-1980s.

No comprehensive academic study of the effectiveness of EU environmental law exists: such a study would be very difficult to carry out due to the sheer volume and variety of the relevant legislation. However, a growing number of studies focusing on particular directives (see e.g. Glim 1990),[1] on individual Member States (Haigh 1990)[2] or on certain sectors of environmental law (Haigh 1986; Bennett 1991),[3] many of them funded by the European Commission (Haigh 1986, 3), have been published. All these studies reveal serious deficiencies in the implementation of Community environmental law in the Member States and highlight the discrepancy that often exists between the formal transposition of directives into national law and their practical application on the ground. In this paper we shall analyse the ongoing academic and policy debate on the effectiveness of European Commission (EC) environmental law. We will discuss the various mechanisms for the enforcement of EC environmental law from a multilevel governance perspective, focusing on the respective roles and sometimes conflicting interests of the various actors involved at the national and EU level.

THE ISSUE OF THE EFFECTIVENESS OF LAW IN THE COMMUNITY LEGAL ORDER

It is generally assumed that the unique features of the Community legal order – which has been described by the Court of Justice as 'a new legal order of international law for the benefit of which the states have limited their sovereign rights' (Case 26/62) – tend to make EC law considerably more effective than classical international law.[4] The supranational nature and autonomous legislative authority of the Community institutions, their power to impose legal obligations on Member States without unanimous consent in a substantial number of policy areas (including to a large extent environmental policy) and the special legislative instruments available to them, whose legal effect is not conditional on national ratification, all concur to endow the European Community with far reaching regulatory powers unmatched by any intergovernmental organization, let alone by conventional treaty making processes. The supervisory and enforcement powers conferred on the Commission, an independent supranational body with an express mandate to ensure the application of Community law, and the existence of a Court of Justice (hereafter, 'the Court') whose jurisdiction is recognized as compulsory by all the Member States and which, since the entry into force of the Treaty of European Union (TEU), can even impose financial penalties on them in certain circumstances, are powerful devices to guarantee the Member States' compliance with their obligations. Last but not least, the well established supremacy of Community law over national law, the doctrine of direct effect, and the Community's elaborate system of judicial protection of individual rights, also have the potential to contribute significantly to the effectiveness of Community law.

The effectiveness of Community law is a multifaceted issue, which can be envisaged from a variety of different viewpoints. In the legal literature, it is most often conceptualized from the individual legal subject's point of view. From this perspective, which is primarily concerned with ensuring the enforcement of individual rights granted by Community law, effectiveness tends to be perceived as essentially a question of judicial protection. In their plea for a strengthening of the effectiveness of Community law through the further development of judicial remedies, published in 1991, on the eve of the signing of the TEU and the Court's *Francovich* ruling (Joined Cases C-6/90 and C-9/90), the editors of the *Common Market Law Review* ('How to Strengthen,' 1991, 713) almost equated effectiveness with judicial protection:

> As far as the effectiveness of Community law is concerned, the single most important instrument for enforcing compliance with Community law continues to be the availability of legal redress open to private individuals in the national courts.

In a similar vein it has been argued that effectiveness is nothing less than 'an overarching principle' of the Community legal order (Weatherill 1995, 116), of which the twin doctrines of supremacy and direct effect are but manifestations. In this view, effectiveness is also in essence a matter of individual rights and remedies:

> Effectiveness encapsulates the notion that Community law shall confer rights plus remedies that make real the practical enjoyment of those rights. (ibid, 117)

The Court of Justice itself has of course recognized the close link between judicial protection and effectiveness, by making the *effet utile* doctrine one of the central tenets of its theory of direct effect. The Court has on several occasions referred explicitly in its judgments to the need to avoid 'a considerable breach in the effectiveness' of Community provisions (Case 65/76, para 20; Schermers and Waelbroeck 1992, 20), or the need not to impair 'the full effectiveness of Community law' (Case C-213/89, para 21). But it would be wrong to reduce the concept of effectiveness to its sole 'effective remedy' dimension.

Quite another perspective on effectiveness is that of the legislator or policy maker, who will be mainly interested in compliance by those to whom norms are directed and, more generally, the achievement of the public policy goals that the law is designed to attain. Of course, where the primary purpose of a law is to establish and protect particular individual rights, both perceptions of effectiveness will largely coincide. But, where the legislator's purpose is not to confer rights but, rather, to regulate certain activities which may produce socially undesirable results by imposing obligations on particular subjects of law and ensuring their compliance with those obligations, a perspective of effectiveness which focuses on judicial protection will not be especially relevant. Since most environmental law is of the latter kind, this study will approach the question of effectiveness from the policy maker's perspective.

There are a number of academic studies assessing the effectiveness of EC law from a policy angle, which can be categorized under the common denominator of 'implementation research' (Siedentopf and Ziller 1988; Schwarze et al. 1994). These studies, mostly in the area of Community economic law, are the result of vast collective research enterprises, involving interdisciplinary teams of researchers in a large number of

Member States, studying the implementation at the national level of particular pieces of Community legislation on the basis of a common research design and methodology, at the interface of law and policy science. The purpose of such studies is to analyse problems of, and obstacles to, implementation of Community law with a view to identifying the conditions of its effectiveness[5] and articulating their implications for the Community's legislative policy. Research of this kind generally takes Community policy and law itself for granted and views effectiveness essentially as a problem of implementation. Its concern is not to develop a theory of effectiveness, but rather to produce practical 'recipes' for policy makers.

One of the practical questions addressed is to what extent the *form* of Community law – the type of legislative instrument used – has an impact on national implementation.[6] A study focusing on the internal market found that neither European Union Directives nor Regulations could be shown to be 'preferable in principle' from the viewpoint of effectiveness (Schwarze et al. 1994). Among the conditions of effective implementation identified, the early involvement in the Community legislative process of the specialized national administrations that will be in charge of transposition and application (Siedentopf and Ziller 1988, 178), as well as timely consultation of the target groups of the legislation (ibid; Schwarz et al. 1994, 92) were found to rank highly. Other important determinants of implementation include the 'acceptance of the Community's objectives by economic operators and Community citizens,' the 'executability and acceptance at national level' and the 'comprehensibility, clarity and coherence of the measures' (ibid, 91).

Another interesting finding is that there is no intrinsic difference between the implementation of Community law and that of domestic law: 'Community law, once it has been incorporated, is applied neither better, nor worse than national law' (Siedentopf and Ziller 1988, 199); it 'suffers from the same shortcomings in implementation and enforcement' (ibid, 176) as are associated with each national legal and administrative culture. On the other hand, one study which compared EC law with European Environment Agency law concluded that a significant difference does exist between Community law and general international law in terms of effectiveness, as 'implementation mechanisms within the Community are considerably more effective than those in the field of application of international treaties' (Schwarze et al. 1994, 93).

THE IMPLEMENTATION PROCESS: A MULTILEVEL REGULATORY CHAIN

Any serious analysis of the problems relating to the effectiveness of Community law requires an understanding of the specific process through which Community secondary legislation is to be implemented in the Member States. This process depends on the type of legislative instrument used by the Community institutions. The process is the simplest for European Union regulations, which are directly applicable in all the Member States. Consequently they do not need to, indeed may not, be transposed into national law. Their implementation will normally, however, require some form of complementary legislative, regulatory or administrative action by the Member States, for example where competent authorities in charge of implementation need to be established or designated by the Member States or where sanctions for non-compliance with the directly applicable provisions of the regulation must be laid down. Nevertheless the chain of intervening steps between legislation and application is rather short and straightforward in the case of regulations.

Most Community legislation in the environmental field is however, in the form of directives, whose implementation is a more complex process. Here the Member States are bound by the result to be achieved, but enjoy considerable discretion as to the means of implementation. The implementation process consists of three distinct steps: *transposition, application and enforcement* (Prechal 1995, 5–6). It is important to distinguish them because there is much terminological imprecision and semantic confusion in the use of these terms,[7] especially in policy documents. It therefore seems preferable to use the term 'implementation' in a general, non-technical sense, to denote 'the entire process by which obligations under Article 189(3) are fulfilled' (ibid, 5), i.e. by which directives are put into effect with a view to achieving the result aimed at. As we shall see this involves more than mere legislative action.

Such action, the first step in the implementation process, is what the concept of 'transposition' refers to, though the terms 'implementation,' 'formal implementation' or 'incorporation' are also often used to denote it (see, e.g. Haigh 1986, 81; Krislov et al. 1986, 59–88). Unlike regulations, directives are not directly applicable and must therefore, first and foremost, be transposed into national law. *Transposition* can thus be defined as 'the process of transforming directives into provisions of national law by the competent national legislative body or bodies'

(Prechal 1995, 5). Or as the Commission (1996a) defined it in its 1996 Communication on the implementation of Community environmental law:

> any legislative, regulatory or administrative binding measure taken by any competent authority of a Member State in order to incorporate into the national legal order the obligations, rights and duties enshrined in Community environmental directives.

Transposition is an obligation to be performed by the Member States within a specified period, determined in every directive. It requires the adoption of provisions which are legally binding within the domestic legal order of each Member State. It is for the constitutional framework of the Member State concerned to determine which authority is responsible for transposition in the domestic legal order. In federal states, subnational authorities may be competent for the transposition of environmental directives.

The next stage in the process of implementation is *application*, sometimes also referred to as 'practical application' (European Commission 1996a) or 'practical implementation' (Haigh 1986, 81). It is not sufficient for the Member States to enact the necessary implementing provisions: they must also ensure that these provisions are effectively applied in concrete cases (Prechal 1995, 6), that the general prescriptions of the directive and its implementing legislation are translated into individual administrative decisions where necessary, and that all other administrative measures required to carry out the provisions of the directive and achieve their purpose are actually taken. The obligations relating to application are normally to be performed by administrative authorities within the Member States, often at various levels of government. Member States must ensure that their competent authorities are provided with the infrastructure and the human and other resources needed to perform these tasks (European Commission 1996a). In the environmental field, the application of directives involves such measures as issuing permits setting emission limit values for individual pollution sources or fixing environmental quality standards for particular areas in accordance with Community law, establishing and implementing emission reduction programs for particular pollutants or sources, gathering certain environmental information and communicating it to the Commission, etc.

Finally enforcement, in so far as it relates to the obligations of Member States, can be defined as 'the process of compelling observance of the directive, either as such or as the national measures transposing it'

(Prechal 1995, 6). Like application it involves measures to be taken by competent authorities of the Member States, but in this case not only by administrative agencies but also by law enforcement authorities and national courts. The purpose of these measures is to compel the individual legal subjects concerned to comply with their obligations under the directive, if necessary through the application of sanctions. This also implies a duty on the part of the Member States to take adequate action to monitor compliance and identify cases of non-compliance, through reporting systems, inspection procedures and similar measures.

As one can readily see from the above overview, the potential causes of ineffectiveness of Community law all along this 'regulatory chain' (European Commission 1996a) are manifold. The most publicized, manifest and straightforward form of ineffectiveness of directives – and also the focus of most Article 226 litigation – is non-transposition or incorrect transposition into national law. But timely and correct transposition is by no means a guarantee of effectiveness. Directives which have been transposed may nevertheless be ineffective due to their inherent lack of normative force, deficient application or inadequate enforcement.

In his typology of situations of non-compliance with Community law, Weiler (1988, 341–2) distinguishes three basic types of non-compliance: non-transposition (including wrongful transposition); non-application (incorrect administration or contravening national measures) and non-enforcement. Depending on the motives of the non-compliant Member State, a further distinction can be made between open and deliberate defiance of Community law, 'benign non-compliance' (i.e. non-compliance which is not deliberate but rather the result of neglect or objective difficulties) and 'evasive non-compliance' (an intermediate category) (Krislov et al. 1986, 63–4).

Many instances of non-compliance with Community environmental law fall into the last category. Indeed in the field of environmental policy, it is not infrequent for Member States to ensure formal transposition of directives, but subsequently fail to take the necessary measures for their application and enforcement. It is much easier to produce 'law on the books' than to actually make it operational. And such cases of evasive, insidious non-compliance with Community law are much more difficult to detect than plain cases of non-transposition. Yet they are not uncommon, as will appear from a number of examples in this study. A Member State may faithfully transpose a provision of a directive that requires discharge permits for a particular sector of industry to be reviewed every four years but abstain from taking any action to ensure that the competent administrative authorities effectively do so. It may explicitly prohibit the emission of certain pollutants but never monitor compliance with this prohibition. This might even be done less by negligence than by design. The lack of any systematic checks by the Commission on practical application and enforcement of directives indeed makes it attractive for Member States to put up a facade of formal compliance.

ENFORCEMENT MECHANISMS: CENTRALIZED AND DECENTRALIZED

Having examined the process of implementation of Community law and the possible dysfunctions in this process, which may result in ineffectiveness, I shall now discuss the specific enforcement mechanisms within the Community legal system designed to ensure the effectiveness of Community law. The EC Treaty provides for both centralized and decentralized enforcement mechanisms (Ehlermann 1988, 145; Prechal 1995, 89), with a key role for the European Commission, the European Court of Justice and the national courts of the Member States. Citizens and civil society organizations, too, have a role to play in enforcing EU environmental law, a role which was actively encouraged by the Commission during the 1980s and 1990s, but is viewed somewhat less positively today, as appears from the most recent Commission policy documents.

The centralized mechanism, which involves both the Commission and the Court, is the one provided by Article 226 of the EC Treaty (formerly Article 169 of the European Economic Community Treaty). Under this provision, the Commission, after a number of preliminary administrative steps, can initiate *infringement proceedings* against Member States before the Court of Justice for any failure to comply with their Community obligations. These proceedings will result in a declaratory judgment of the Court. Though Article 226 proceedings can be used to challenge any form of non-compliance by a Member State with Community law, in practice the vast majority of cases brought under this provision concern non-transposition or incorrect transposition of directives. The Commission has so far concentrated its enforcement action on this most obvious and straightforward form of non-compliance, though a recent shift towards increased attention to problems of inadequate application and enforcement at the national level has been noted by commentators (Prechal 1995, 7–8, 58–60). The problem of non-application has been likened to the proverbial iceberg, 'the tip of which we have seen but the bulk of which is so far undetectable' (Krislov et al. 1986, 73). In 1993, the TEU granted the Commission and the Court a new and powerful tool to enforce Community law – an instrument wholly unprecedented in international law – the power to impose financial penalties on delinquent Member States, under Article 228(2) EC.

Article 226 however, is only one of the instruments used by the Commission to secure the effectiveness of Community law. The attention devoted to this procedure in most textbooks tends to obscure the fact that the Commission in practice principally uses *non-legal* means for this purpose. To quote Francis Snyder (1993, 48), the author of a rare study that considers the issue from a more theoretical perspective, 'the effectiveness of Community law, as its making, depends primarily on negotiation.' He argues that the different instruments used by the Commission, including Article 226 proceedings, can in fact be viewed as part of what he calls an overall 'process of administrative negotiation' (Snyder 1993, 27). The fact that such proceedings are usually terminated during their administrative stage and only rarely reach the Court of Justice, indicates that 'the main form of dispute settlement used by the Commission is negotiation, and litigation is simply a part, sometimes inevitable but nevertheless generally a minor part, of this process' (Snyder 1993, 30). Since the Commission enjoys total discretion in the use it makes of Article 226, the threat of litigation is often used as a bargaining chip in the negotiation process.

According to Snyder (1993), a non-legal technique which is increasingly being used by the Commission to promote the effectiveness of Community law is, somewhat paradoxically, soft law. Non-binding instruments such as Commission Communications are used as a complement to hard law to bypass the Council and 'provide guidelines for negotiating the effectiveness of Community law,' for example by interpreting the broader legal and policy implications of important judgments of the Court of Justice (Snyder 1993, 33). Finally the Commission also relies on what Snyder calls 'structural reform,' as yet another technique to ensure the effectiveness of Community law: it involves establishing various forms of administrative cooperation and networking between the Commission services and national or subnational administrations.[8]

Though Snyder's analysis could be criticized for somewhat overstretching the notion of effectiveness of Community *law* as opposed to *policies*, it is useful for putting the role of Article 226 proceedings in a proper perspective. I would venture to submit that the Commission's preference for negotiation as an 'enforcement' strategy may also be a reflection of the prevalence of normative ambiguity in Community law and its concern to remain in control of interpretation for reasons of political expediency. When a dispute with a Member State is brought before the Court of Justice, the Commission relinquishes control to the Court. Since the Commission is not primarily concerned with the vindication of particular individual rights but rather with the overall effectiveness and integrity of Community policies, its purposes may frequently be better served by negotiated interpretation of Community law than by authoritative but often unpredictable judicial interpretation. Thus the negotiation process that precedes the adoption of secondary legislation does not really end at that time but carries on, as it were, into the implementation stage.[9]

There are of course also practical limits to recourse to the rather cumbersome and time-consuming medium of infringement proceedings. Senior Commission officials have referred to 'the unsuitable character of the infringement procedure for correcting specific cases of wrongly-implemented Community law' (Ehlermann 1988, 146) and, consequently, argued that:

> instead of concentrating only on the functioning of centralized control, the Commission, the European Parliament and the Council should focus as much, if not more, attention on strengthening decentralized control of the implementation of Community law. (Ehlermann 1988, 148)

While it is obviously impossible, due to time and resource constraints on both the Commission and the Court, to deal with all individual cases of non-application or improper application of Community law through this procedure, there is no reason why, in principle, it should not be used in such individual cases, which are often exemplary and raise important legal issues. But at any rate, it is clear that a decentralized enforcement mechanism serves a necessary complementary function to the centralized mechanism of Article 226.

The notion of decentralized enforcement refers to judicial redress for violations of Community rights available to individuals in national courts, which may call on the Court of Justice to give preliminary rulings on the interpretation of relevant provisions of EC law under Article 234 (formerly Article 177) of the Treaty. The Court has made use of this opportunity to gradually expand the nature and scope of judicial protection available through national courts, based on the doctrine of the primacy of Community law and a number of related doctrinal constructs – such as direct effect, *effet utile*, Community loyalty, consistent interpretation of domestic law, state liability – designed to provide individuals with effective judicial remedies for the enforcement of Community rights through the domestic court systems of the Member States.

Though it may have been 'elaborated … less by design than in an *ad hoc*, reactive and unsystematic manner' (Snyder 1993, 40), this decentralized enforcement system is unique to the Community legal order and has undoubtedly made a significant contribution to the effectiveness of Community law in a number of areas, especially in the field of economic law. Judicial policy making through adjudication is of course reactive by its very nature, as the Court of Justice has no control over the cases that are submitted to it and thus cannot determine its own agenda (Snyder 1993, 50). As the decentralized enforcement system is triggered exclusively by *ad hoc* individual claims, its adequacy in ensuring the overall effectiveness of Community law is questionable (Snyder 1993, 51).[10] Its success is dependent on the existence of well-established individual rights and strong private interests in their enforcement, which is not a typical feature of EC environmental law.[11]

ENFORCEMENT AND EFFECTIVENESS OF COMMUNITY LAW AS A POLICY ISSUE

In recent years the enforcement and effectiveness of Community law is increasingly being perceived as a fundamental policy issue in most areas of EC legislative activity. This new political concern for the effectiveness of EC law is part of a broader public debate on its adequacy, quality and accessibility, which developed in the wake of the signing of the Maastricht Treaty (European Council 1993, 1). In a declaration 'on the implementation of Community law' annexed to the TEU, the Maastricht Intergovernmental Conference recognized the importance of full and accurate transposition and effective enforcement of Community law in the Member States. Specifically with respect to enforcement, the declaration stressed that it:

> is essential for the proper functioning of the Community that the measures being taken by the Member States should result in Community law being applied with the same effectiveness and rigour as in the application of their national law. (Final Act 1992, 231)

In a 1996 Communication on 'The impact and effectiveness of the internal market' (European Commission 1996b), reviewing the effectiveness of the Community legislation adopted pursuant to its 1985 White Paper on 'Completing the Internal Market' (European Commission 1985), the Commission found that there were still important deficiencies in the implementation and enforcement of Community law in this field and concluded that proper enforcement in the Member States is 'a growing challenge' which requires more attention at the political level. The Communication proposes an 'eight-point programme' of actions to improve enforcement at both the national and Community level, including such measures as accelerating Commission action on infringements, carrying out Community audits of national enforcement systems and policies and promoting the involvement of national courts in the enforcement of Community law (ibid, Section 5.1, 30–31). The same year, the Commission released another Communication (European Commission 1996a) proposing very similar measures to improve the implementation of Community law in the field of the environment. This will be further discussed below.

Whether such legal, procedural and institutional measures can substantially improve the effectiveness of Community law seems rather uncertain, as they do not address the *substantive* content of EC law and the *political context* in which it is made. Joseph Weiler, commenting in 1995 on the Commission's 1985 White Paper, in which it already announced its intent 'to continue improving and rationalizing its internal procedures in order to correct violations [of Community law] rapidly and effectively' (European Commission 1985, para 154), criticized the Commission for failing 'to deal adequately with the complexity of the [non-compliance] phenomenon' (Weiler 1988, 338). In his diagnosis of what he calls 'the pathology of non-compliance' affecting the Community (Weiler 1988, 340), Weiler has highlighted its paradoxical nature: the fact that notwithstanding the dominant control of the Member States over the Community legislative process, which effectively allows them to block any politically unacceptable legislation, these very same Member States have become increasingly delinquent in their application of Community law. He has identified a variety of political factors in decision making processes at the Community and Member State level which explain this 'non-compliance paradox' (Weiler 1988, 346–56). The 'cost of compromise' features prominently among those factors (Weiler 1988, 353–4). In Weiler's assessment (1988, 357), non-compliance 'is conditioned already by the decisional process, is aggravated by the type of legislative measures adopted, and is not amenable to centralized Commission [enforcement] action.' Francis Snyder, too, has stressed the importance of extra-legal factors in determining the effectiveness of Community law. In his view this effectiveness depends not only on institutions, but also on 'corporatist arrangements, pluralist politics, policy networks and individuals' (Snyder 1993, 19–20).

THE POLICY DEBATE ON THE EFFECTIVENESS OF EC ENVIRONMENTAL LAW

The apparent lack of effectiveness of Community environmental law became a political issue in 1984, when the European Parliament undertook an enquiry into the implementation of the EC waste management legislation and adopted a resolution criticizing the Member States for their failure to implement this legislation fully and correctly and the Commission for not having closely supervised its implementation (European Parliament 1984, 67). As a result, the Commission took internal administrative measures in 1984 to ensure more effective supervision and enforcement of Community environmental law by its services (Krämer 1988, 23). The second half of the 1980s saw a steep rise in the number of infringement proceedings for non-implementation of environmental directives brought by the Commission against Member States under what was then Article 189 of the European Economic Community Treaty (see Collins and Earnshaw 1993, Table 1, 216).

In its Fourth Environmental Action Programme submitted to the Council in October 1986, the Commission for the first time devoted a full chapter to the implementation of Community directives, in which it recognized that not only transposition into national law, but also 'practical implementation at the national level of the requirements of Community legislation and its effectiveness in reality in improving the quality of the environment ... raise considerable problems' (European Commission 1986, para 2.2.4). Affirming that it would henceforth consider 'complete and effective implementation' of EC environmental law as 'a matter of priority concern' (European Commission 1986, para 2.2.8), the Commission announced a number of specific policy measures, complementary to legal action under Article 189, designed to promote better implementation. These include rather 'soft' actions, like a more intensive 'dialogue' with competent administrations in the Member States (European Commission 1986, para 2.2.6) and increased publicity and training activities. But the Commission also indicated that it would encourage citizens, non-governmental organizations and local authorities to make use of their right of complaint to the Commission and would also examine whether 'Community environment inspectors should be appointed to work with national officials' (European Commission 1986, para 2.2.7). The number of citizen complaints about cases of non-compliance with EC environmental legislation increased significantly from 1986 to 1990 (see Collins and Earnshaw 1993, Table 5, 232) but the idea of a Community environmental inspectorate, though a recurring theme in the policy debate, has not materialized so far.

In 1990 the importance of implementation and enforcement was acknowledged at the highest political level, when the European Council, in its Dublin Declaration on 'The Environmental Imperative,' stressed that 'Community environmental legislation will only be effective if it is fully implemented and enforced by Member States' (European Council 1990, 18-21). With an undeniable sense of irony in view of most Member States' poor implementation record, the heads of state and government, in their declaration, 'renewed their commitment in this respect.' Two years later, the Court of Auditors added its grist to the mill in a highly critical report on Community environmental policy, noting the 'significant gap between the set of rules in force and their actual application' (Court of Auditors 1992, 19).[12]

Thus it comes as no surprise that implementation and enforcement of Community environmental law again featured as a prominent issue in the Fifth Environmental Action Programme, though the context in which it was addressed by the Commission had changed. While the Fourth Programme did not directly question the adequacy of the Community legislation itself but only the adequacy of its implementation at Member State level, the overall tone of the Fifth Programme is less self confident. One of its main political messages is that Community environmental policy so far has relied too heavily on a 'top-down' legislative approach (European Commission 1992, Vol II, 9, para 34) and needs 'a broader mix of instruments' (ibid, 8, para 31). On the other hand, the full chapter devoted to 'Implementation and Enforcement' (ibid, Chapter 9, 75–7) indicates that this note of scepticism about the formerly prevailing legislative strategy should not be taken as a signal that the Commission no longer intends to pursue enforcement of existing Community legislation. Apart from reforms to improve the quality of future legislation and ensure its full integration with other policy instruments, 'stricter compliance checking and enforcement' is also one of the goals laid down in the Fifth Action Programme (ibid, 75).

More specifically, the Commission calls upon Member States to transpose directives into domestic law, put in place the other implementing measures required and report to the Commission within the timeframes laid down by the Council,[13] and also urges them to 'establish enforcement bodies and procedures' at the national level 'to ensure full and equal compliance with both legislative and authorization-associated requirements' (ibid, 76). As in the Fourth Action Programme, much reliance is placed on the role of citizens: the Commission recommends that 'an accessible and efficient complaints facility should be developed at local, regional and national level' to handle complaints from the public, which 'should be considered less a nuisance than a resource' (ibid), and further calls for 'practicable access to the courts' to be granted to citizens and public interest groups at the national level 'in order to ensure that their legitimate interests are protected and that prescribed environmental measures are effectively enforced and illegal practices stopped' (ibid, 76–7). Thus the Fifth Action Programme places the ball squarely in the Member States' court.

THE 1996 COMMISSION COMMUNICATION ON IMPLEMENTING COMMUNITY ENVIRONMENTAL LAW

Following the endorsement of the Fifth Environmental Action Programme by the Council in 1993, the topic of implementation remained on the Community's political agenda. The Commission's intention to keep it high on that agenda is apparent from the initiative it took to further articulate its policy on the issue in a Communication to the Council and the European Parliament. This Communication entitled 'Implementing Community Environmental Law' (European Commission 1996a), was published in October 1996. The stated purpose of the Communication is 'to increase the effectiveness and efficiency of Community environmental law with consequent improvements for the European environment' (ibid, 7, para 19), both by 'reinforcing' existing mechanisms and by proposing new areas for action. It is important to note that the communication does not focus exclusively on enforcement action by the Commission, nor even on Community action in general, but in line with the Fifth Action Programme's emphasis on 'shared responsibility' (European Commission 1992, Vol II, Chapter 8), covers measures to be taken by all the actors in the 'regulatory chain' and at all stages of this chain. Although the Commission stresses that it 'intends to continue to make full use of its enforcement powers based on Article 169' (European Commission, 1996a 4, para 10), the overall effect of this approach, whether intended or not, is to downplay the importance of the specific legal mechanisms provided by the Treaty to ensure effective implementation of Community law. Indeed the Communication itself advocates the use of alternative mechanisms precisely by emphasizing that they 'will strengthen enforcement and, at the same time, *ease the control of Member States by the Commission*' (ibid, 8, para 25 (emphasis added)).

What then are these new mechanisms that the Commission intends to promote? The importance of adequate compliance monitoring and enforcement of environmental requirements by competent authorities in the Member States, which had already been stressed in the Fourth and Fifth Action Programmes, is again emphasized. But the earlier idea of appointing Community environmental inspectors, first mentioned in the Fourth Programme, is presented in a very watered down form. The Communication goes out of its way to reassure Member States that inspection on the ground is their task and that it is up to each Member State to choose the appropriate means for fulfilling it. The Commission merely intends to 'consider making recommendations in order to assist Member States in carrying out inspection tasks, by the establishment of guidelines' (ibid, 9, para 29). Later, if the practice under these guidelines indicates that further action is needed to ensure even application of environmental law throughout the Community, the Commission may consider proposing the establishment at Community level, not of an inspectorate, but of what it describes as 'a limited body with auditing competences' which could supervise and 'audit' how Member States fulfil their inspection tasks (presumably on the basis of the guidelines laid down by the Commission) (ibid).

The other two new measures proposed in the Communication are an elaboration of recommendations contained in the Fifth Action Programme: the establishment of mechanisms at the national level to handle complaints by citizens and the broadening of access to justice in environmental matters. Recognizing the limitations of the enforcement mechanisms available at Community level in dealing with problems 'in relation to the practical application and enforcement of Community environmental legislation by public authorities' (ibid, 10, para 32), the Commission calls on Member States to institute appropriate procedures for this purpose at their level. The Communication is rather vague as to the exact nature of these national procedures: it describes them as 'non-judicial' (ibid, para 31) and refers to the 'ombudsman' systems existing in some Member States as an example of what might be envisaged (ibid, 11, para 34). However, the Commission announces its intention to consider recommending 'minimum criteria for the handling of complaints and carrying out of environmental investigations in Member States' (ibid, para 35). Finally, recognizing that 'the public and public interest groups do not as a general rule have sufficient access to the national courts of the Member States in environmental matters' (ibid, 12, para 39) and that opening up such access would contribute to the enforcement of Community environmental law, the Commission will 'examine the need for guidelines' on access to national courts by representative non-governmental organizations (ibid, 13, para 43).

It is striking that this new 'action' envisaged by the Commission to increase the effectiveness of Community environmental law involves recourse only to *soft law* measures (recommendations or 'guidelines') at Community level. In its political response to the Commission Communication, the Council has eagerly endorsed this soft approach and made it quite clear that there is little support for more far reaching measures of a legal or institutional nature. Thus on the issue of inspection, the Council resolution stresses that the different inspection practices in the Member States 'should not be replaced by a system of inspection at Community level' (European Council, 1997, 1, para 16) and implicitly though unambiguously rejects even the Commission's modest idea of some sort of 'auditing body' at Community level. Indeed the Council asks that not only the 'minimum criteria and/or guidelines for inspection tasks' contemplated by the Commission, but also 'the possible ways in which their application in practice could be monitored *by Member States*,' be submitted 'for further consideration in Council' (ibid, para 17 (emphasis added)). As to access to justice in the Member States, the resolution invites the Commission to submit to the Council a report on the existing situation in the Member States before assessing '*whether there is a need for* the development of minimum criteria or guidelines' in this field (ibid, para 25 (emphasis added)).

THE ACADEMIC LITERATURE ON THE EFFECTIVENESS OF COMMUNITY ENVIRONMENTAL LAW

As in the area of international environmental law, much of the literature on the effectiveness of Community environmental law has focused on *enforcement* problems.[14] Most attention has been devoted to the role of the Commission in monitoring compliance by Member States and in making use of its power to take legal action against them under Article 226 of the Treaty (see Krämer 1988; Krämer 1992, 209–28; Krämer 1995, 135–44; Wägenbaur 1991; Macrory 1992; Macrory, 1996; Reitzes 1992). The limitations to the use of this procedure as an instrument for the enforcement of Community environmental law, especially to deal with the hundreds of complaints about practical non-compliance in the Member States, have been duly recognized. They result not only from the staff and resource constraints of the Commission (Krämer 1992, 219), from the lack of adequate reporting by the Member States,[15] from the fact that the procedure is targeted at central governments while many infringements are actually committed by subnational authorities (Macrory 1992, 357–8), and from the absence of any preventive effect due to the long duration of infringement proceedings, but also from the political context in which the Commission fulfils its role as 'guardian' of Community law. It has indeed been pointed out that conflicts of interests may arise between the Commission's enforcement function and its political role in the Community's legislative process (Williams 1994, 351–99. See also Krislov, Ehlermann and Weiler 1986, 87–88).

The recognition of these limitations has led both to proposals for improving the effectiveness of the Article 226 procedure and to a search for alternative means of enforcement. Thus, for example, encouraging the use by individual citizens and non-governmental organizations of their right of complaint to the Commission was advocated as a means of overcoming the inability of the Commission services to systematically monitor the practical application of Community law. On the other hand, the complaint-based enforcement system has been criticized because of its essentially contingent and *ad hoc* nature.[16] Another much debated reform is the establishment of some form of environmental inspectorate at Community level.[17] While such a measure would no doubt increase the Commission's independent monitoring capacity, it is not only politically unfeasible – as the Council's recent reaction to even the Commission's minimalist proposal for an 'auditing body' indicates – but also intrinsically incapable of remedying the other deficiencies of the Article 226 procedure, which would remain the only means available to the Commission to take enforcement action based on its inspectors' findings.

Some authors have explored the potential for use of the direct effect doctrine as an instrument for the enforcement of Community environmental law through national courts and referrals to the Court of Justice under Article 234 of the Treaty (Jadot 1987; Jadot 1989; Krämer 1990b; Krämer 1991; Krämer 1996; Pagh 1995). But while this doctrine has undeniably contributed significantly to enhanced judicial protection of individual rights and to the effectiveness of Community law in other fields (see generally, e.g. Curtin 1990; Plaza Martin 1994), its use is dependent on effective access to national courts and the existence of strong private interests in securing enforcement. In the environmental field however, as the Commission itself has observed in its recent Communication, there are serious legal obstacles to public interest litigation by non-governmental organizations in most Member States. As to the 'decentralization' of enforcement through the establishment of 'non-judicial' procedures at the Member State level to handle complaints from citizens about infringements of Community environmental law, as advocated by the Commission in the Fifth Environmental Action Programme and in its 1996 Communication,[18] it is hard to see exactly how such a system of 'self policing' could be adequate to deal with non-compliance *by national authorities themselves*.

An important development since the entry into force of the TEU in 1993 and one which may have a significant impact on the effectiveness of the Commission's enforcement efforts, is the power of the Court of Justice, under Article 228(2) EC, to impose 'a lump sum or penalty payment' on a Member State which has failed to comply with an earlier judgment of the Court. The Court's ability to exercise this power depends on the Commission's initiative. It is up to the Commission to bring such cases before the Court and to 'specify' the amount of the financial penalty 'which it considers appropriate in the circumstances.' In its Communication on the implementation of Community environmental law, the Commission indicated that it intends to make 'full use' of this possibility to ensure Member States' compliance with their environmental obligations (European Commission 1996a, 4, para 11).

However, it took the Commission a few years to lay down its general policy with respect to the use of this new legal instrument. In June 1996 it adopted criteria for the application of Article 228(2) and determined that penalty payments were the 'most appropriate instrument' to bring about compliance.[19] Half a year later, the Commission agreed on a method for calculating the amount of the penalty payments it would ask the Court to impose, based on the seriousness and duration of the infringement and each Member State's financial capacity (European Commission, 1997a). A few weeks after this decision, the Commission first requested the Court to impose penalty payments against three Member States in a number of environmental cases.[20] The potential effectiveness of the financial sanctions of Article 228(2) is, however, limited by the fact that recourse to them is just as discretionary on the part of the Commission as the initial Article 226 infringement proceedings themselves. Moreover, the procedure can only be used in a second action, after an initial ruling under Article 226 has been ignored, which means that it is only several years after the original infringement has been discovered that financial penalties could effectively be imposed.

DIMENSIONS OF EFFECTIVENESS: ENFORCEMENT AND NORMATIVE QUALITY

The rather one sided emphasis of much of the literature and political debate on enforcement mechanisms has tended to obscure the other dimensions of the problem of the effectiveness of Community environmental law. The question which should logically be the first and most obvious is rarely explicitly addressed: *what legal prescriptions are there to enforce and how adequate are they in the first place?* The fundamental question of the normative quality of EC environmental law has been noted in passing by some commentators (see e.g. Krämer 1995, 142), but has, to my knowledge, been the subject of only a single article length study (Jans 1990)[21] so far. And, since the publication of the Fifth Environmental Action Programme, it is only recently that this issue has also surfaced in the policy debate.

In the Fifth Action Programme's chapter on implementation and enforcement of Community environmental law, the Commission admitted that the effectiveness of implementation did not depend only on enforcement procedures, but was also 'closely related to *the quality of the measures themselves*' (European Commission 1992, Vol II, 75 (emphasis added)). The chapter therefore starts off with a diagnosis of what the Commission considers to be the reasons for the implementation problems that have arisen. The very use of legislation as the main instrument of the Community's environmental policy is called into question, where the Commission argues that 'perhaps too great a reliance was placed on regulation of the 'command and control' type' (ibid), while 'the legislative approach may not always be the best choice as a first step' (ibid, 76). But the choice of legislative instrument – the 'preponderant recourse to Directives' rather than regulations – and the deficiencies of the legislative process itself – the need for political compromise in Council – are also blamed as causes of ineffectiveness (ibid, 75).

The Commission's 1996 Communication on implementation, in which this analysis is further elaborated, contains a section entitled 'quality of Community legislation' (European Commission 1996a, 14–16), which focuses on how to improve the drafting of legislation to avoid 'ambiguous, unclear or complicated provisions' which 'lead to incomplete or uneven implementation throughout the Community' (ibid, 14, para 45). The Communication of course does not give the full picture where it asserts, glossing over the internal negotiation process *within* the Commission (see Krämer 1992, 139–40) that 'unclear and ambiguous drafting of environmental legislation is very often the result of compromises needed *at the Council level* or following the readings of the European Parliament' (European Commission 1996a, 14–15, para 46 (emphasis added)). At any rate, the solutions proposed in the Fifth Action Programme and in the subsequent Communication, designed to improve the 'transparency' of the legislative process[22] through various kinds of additional consultation mechanisms in the drafting stage, do not seem likely to be very effective. Indeed, normative ambiguity is an essential structural feature of Community environmental legislation and an unavoidable consequence of a law making process still primarily based on intergovernmental political bargaining, as the Council itself implicitly admitted in calling for 'increased efforts ... to ensure that texts resulting from the different phases of the Community legislative process are not *unnecessarily* ambiguous or complicated' (European Council 1997, para 8 (emphasis added)), in spite of its earlier, general exhortation that the wording of Community legislative acts should be 'unambiguous' (European Council 1993, 1, para 1).

THE DEBATE ON IMPLEMENTATION SINCE THE SIXTH ENVIRONMENTAL ACTION PROGRAMME

According to the preamble to the Sixth Environmental Action Programme (6EAP), 'full and correct implementation of the existing legislation is a priority.' Accordingly one of the 'strategic approaches' of the Programme, as listed in Article 3 of the 2002 Decision of the European Parliament and Council approving it, is 'encouraging more effective implementation and enforcement of Community legislation on the environment.' This is to be achieved *inter alia*, by 'increased measures to improve respect for Community rules on the protection of the environment and addressing infringements of environmental legislation.' Such measures are first and foremost, the responsibility of the Member States, but where they fail in their duty to ensure adequate transposition, implementation and enforcement of EU environmental legislation, the powers of the Commission under the Treaty to monitor national implementation and initiate infringement proceedings against delinquent Member States come into play. Though not explicitly mentioned in the 6EAP, the harmonization of national provisions concerning sanctions for non-compliance with environmental law and access to administrative and judicial review procedures in the Member States are also relevant in this context.

As the Commission points out, implementation and enforcement of EU environmental law in each individual Member State is of crucial importance to the effectiveness of environmental policy and this final link in the 'regulatory chain' still leaves much to be desired: 'The relatively high number of complaints received by the Commission reflects the non-existence and/or the relative lack of efficiency of complaint mechanisms in Member States' (European Commission 2003a, 35). Where competent authorities in the Member States fail to enforce or even deliberately violate EU environmental legislation, only access to adequate administrative and judicial review procedures at the national level can provide citizens, businesses and non-governmental organizations (NGOs) with means to compel their authorities to comply with their obligations under Community law. Providing such access is an obligation incumbent on Member States under the Aarhus Convention, which has been ratified by the Community in 2005 and thus has become part of EU law.

The Commission deserves credit for having submitted a proposal for a Directive on access to justice in environmental matters (European Commission 2003b) as part of its package of measures for the implementation of the Aarhus Convention, notwithstanding the lack of a clear mandate for such a proposal in the 6EAP. This proposal goes some way towards addressing the issue of improving access to justice in the Member States, but is unlikely to be adopted by the Council any time soon, despite a favourable first reading opinion by the European Parliament in 2004. A 2007 study ordered by the Commission confirms the persistence of serious deficiencies in access to environmental justice in many Member States, despite their ratification of the Aarhus Convention, and strengthens the case for adoption of EU legislation in this field.

Other measures provided for by the 6EAP to improve implementation and enforcement of EC environmental law at Member State level are the promotion of improved standards of permitting, inspection, monitoring and enforcement and improved exchange of information on best practice in this field. A Recommendation of the European Parliament and of the Council providing for minimum criteria for environmental inspections in the Member States (2001/331/EC) was adopted prior to the 6EAP and its implementation is being promoted through the work of the European Network for the Implementation and Enforcement of Environmental Law (IMPEL). In a Communication reviewing the implementation of this Recommendation, the Commission reported 'that there are still large disparities in the way environmental inspections are being carried out in the Member States' despite the Recommendation (European Commission 2007a). The European Parliament in a resolution of 20 November 2008 responding to the Commission's findings, stressed 'that good and even enforcement of Community environmental law is essential, and that anything less falls short of public expectations'. It called on the Commission to come forward with a proposal for a directive on environmental inspection, rather than address the issue through the insertion of specific inspection requirements in sectoral legislation, as the Commission is proposing (European Parliament 2008).

Another important Commission legislative initiative designed to promote better implementation and enforcement of EU environmental law in the Member States is a 2008 Directive on the protection of the environment through criminal law (European Commission 2007b; European Commission 2008a), which lays down some basic harmonized standards at EU level for the application by Member States of criminal sanctions to serious violations of environmental law. The Commission considers that the existing sanctions in the Member States are insufficient to ensure effective implementation of EC environmental policy, due to disparities in the type and level of sanctions. Some Member States provide for administrative sanctions only, whereas the Commission is of the opinion that criminal sanctions should be introduced for all serious environmental offences. The Directive defines those offences and requires Member States to ensure that they constitute criminal offences under their national law. Moreover it would also achieve some measure of harmonization of the level of the sanctions for the most serious offences, by specifying that these should be punishable by imprisonment and laying down its minimum and maximum duration, depending on the nature and circumstances of the offence.

This initiative is the latest development in a long-standing legal and political struggle between the Commission and the Council on the competence of the EC to legislate on environmental criminal law. In the important judgment of September 2005, the Court of Justice held that the Community legislature has the power, under Article 175 of the EC Treaty, to take 'measures which relate to the criminal law of the Member States which it considers necessary in order to ensure that the rules which it lays down on environmental protection are fully effective' (Judgment of 13 September 2005, Case C-176/03).

THE ROLE OF CITIZEN COMPLAINTS AGAINST INFRINGEMENTS OF EC ENVIRONMENTAL LAW

At the European Parliament's request, the Commission publishes an annual report on its efforts to monitor the application of Community law in all areas. Since 1999 the Commission has also issued an annual survey specifically on the implementation and enforcement of Community environmental law. These reports make interesting but puzzling reading for anyone interested in the effectiveness of EC environmental directives and the activities of the EU institutions to compel Member States to comply with their obligations.

In its report for 2002, published in 2003, the Commission noted: 'The last five years have seen a growing difficulty in the timely and correct implementation as well as proper practical application of EC environmental legislation. This is reflected in the growing number of complaints received and infringement cases opened by the Commission every year' (European Commission 2003c, 33). But two years later the annual report stated: 'In the area of the environment, implementation of Community legislation by the Member States has improved in recent years. This is borne out by the substantial reduction in the number of new complaints registered by the Commission in 2004 (336 as compared with 555 in 2002)' (European Commission 2005, 6) According to the latest published data, the number of complaints continues to decline, plummeting to 275 in 2005.

While there have indeed been a falling number of complaints, is it justified to infer from this statistical fact, as the Commission appears to be doing, that there has been an almost miraculous reversal of the situation in terms of implementation of environmental law over two years? Not exactly, because the Commission's own interpretation of the data seems more cautious in the latest annual survey on the implementation and enforcement of environmental law, which points out that 'a further major reduction in the number of open cases is not likely in the near future. This is due to the fact that following enlargement the number of complaints and petitions from the European Parliament concerning breaches of environmental law in the new Member States, including cases related to EU-funded infrastructure projects, is likely to grow significantly' (European Commission 2006, 5).

What then is the situation with respect to the implementation of environmental law in the Member States and the Commission's policy to deal with complaints lodged by citizens?

From the outset it should be pointed out that complaints are only one of several ways in which the Commission detects potential infringements of Community law. It also routinely launches infringement proceedings when Member States fail to notify measures taken to transpose directives or when its own investigations reveal that national legislation is not in conformity with Community law (e.g. as a result of 'conformity checks' carried out by the Commission itself). In the field of environmental law however, complaints still account for more than half of the cases opened every year. Thus, although the Commission's policy in the area of implementation monitoring and enforcement cannot be reduced to its handling of individual complaints only, this question is most directly of concern to the general public. As the European Parliament stressed in a resolution adopted on 16 May 2006, 'the number of complaints relating to infringements of Community law shows that European citizens play a vital role in its application, and that the capacity properly to address their concerns is important for the credibility of the European Union'.

According to published data, the number of new complaints registered by the Commission in environmental matters peaked in 2001, declined slowly until 2003, very significantly in 2004 and again, but more slowly, in 2005. However, the number of complaints relating to the environment is still the highest of any area of EU policy (24 per cent of the total number of complaints, albeit compared to 38 per cent in 2002). Does the declining number of environmental complaints reflect an improved implementation record or rather rising public scepticism about the legitimacy of European institutions in general and the effectiveness of the Commission's role as 'guardian of the Treaty' in particular?

In the past, the Commission has actively encouraged citizens to raise complaints about infringements of EC environmental law. It continues to stress that such complaints 'constitute a vital means of detecting infringements of Community law,' which 'helps to make the principle of a Community based on the rule of law a tangible reality for Europe's citizens and economic operators.' At the same time however, it also emphasizes the limited resources available to the Commission services for handling individual complaints and the 'practical difficulties' in doing so, and admits that it 'therefore concentrates on problems of communication and conformity' of national transposition measures rather than individual cases of bad application of EC environmental law.

In practice, it appears that the proportion of environmental cases that eventually lead to a 'reasoned opinion' being issued against a Member State, let alone to proceedings before the European Court of Justice, is small. In 2005 only some 16 per cent of pending environmental cases (whether detected as a result of complaints or of the Commission services' own monitoring activities) were referred to Court. Thus the Commission's actual practice in the field of environmental law enforcement sends a mixed signal to citizens.

The Commission in fact admits that it is 'particularly common' for environmental cases to be settled before the final stage of proceedings and explains that this is due to the fact that 'a large number of situations to which the Commission's attention is drawn by complaints ... turn out not to be infringement situations as there is no legal basis in Community law or the allegation by the complainants ... is unfounded in fact or in law.' It should be recalled here that the Court of Justice has consistently held that the Commission's decision whether or not to initiate and pursue infringement proceedings against a Member State is fully discretionary.

The important question is however, how the Commission uses its enforcement discretion in environmental cases and on what criteria it bases its decisions not to proceed against a Member State. A survey sheds some further, and rather disturbing, light on Directorate General Environment's policy in this regard.

In response to criticisms from the European Ombudsman about the lack of transparency of such decisions, the Commission in 2002 published internal procedural guidelines concerning relations with complainants in the handling of complaints (European Commission, 2002a), followed by a Communication on 'Better Monitoring of the Application of Community Law' (European Commission 2002b), in which it first specified the 'priority criteria' it would apply in the exercise of its discretion.

These criteria unfortunately, seem to exclude most cases of environmental non-compliance reported by citizens from priority attention: only 'damage to the environment with implications for human health' and 'cases of systematic incorrect application detected by a series of separate complaints by individuals' with respect to the same piece of legislation, or cases of failure to transpose or incorrect transposition which affect a large segment of the public, would fall within the criteria. These general criteria are not fully consistent with earlier statements of enforcement policy, for example in DG Environment's *Annual Survey for 2000–2001* (European Commission 2002c), according to which, individual complaints relating to instances of incorrect application of Community environmental law 'which highlight questions of principle or general interest or administrative practices that contravene the directives' would also be pursued. Has this policy now effectively been superseded by the new, general 'priority criteria' which imply a higher threshold?

As the Annual Survey does not contain any statistics on the Commission's handling of individual complaints about bad application, it is impossible to tell whether there have been any changes in relevant practice in the environmental field since 2002. It is striking however, that there was a significant increase in the overall number of infringement cases closed by the Commission before the reasoned opinion stage (a 20 per cent increase from 2003 to 2004), but since only aggregate statistical data are published in the Annual Report, we cannot determine whether environmental cases were particularly affected by this development.

On several occasions, the Commission has pointed out that, as a result of enlargement, the workload on its services to monitor the implementation of Community law in all Member States and take enforcement action as required would increase substantially. The mere time and effort required to scrutinize the conformity of 25 different national transposition measures in 20 languages needs little explanation. In its Communication of 2002 on 'Better Monitoring', the Commission recognized that more resources would have to be devoted to this task (European Commission, 2002b). Without additional resources, it is doubtful that the objectives of the 6EAP can be achieved and rather more likely that the quality of monitoring and enforcement will decline.

The criteria and resources for dealing with registered complaints are one cause of concern; another are recent allegations that, in the environmental sector, certain complaints by individuals are no longer registered by Commission services, in conflict with official Commission policy as stated in the 2002 Communication to the European Parliament and the Ombudsman. Non-registration of complaints is mentioned as an issue in a report adopted unanimously by the European Parliament's Committee on Legal Affairs, which also refers to ongoing investigations by the Ombudsman on the matter.

THE 2008 COMMISSION COMMUNICATION ON IMPLEMENTING EC ENVIRONMENTAL LAW

On 18 November 2008, the Commission published a new 'Communication on Implementing European Community Environmental Law' (European Commission 2008b), the first comprehensive statement of its policy in this field since 1996. This initiative had been announced in the mid-term review of the Sixth Environmental Action Programme in April 2007 but had been delayed due to disagreements between Commission services.

In the above mentioned general Communication on 'Better Monitoring of the Application of Community Law', published in 2002 (European Commission 2002b), the Commission first specified the 'priority criteria' it would apply in the exercise of its discretion on how to deal with infringements of Community law brought to its attention by citizen complaints. The general criteria for 'prioritisation and acceleration in infringements management' [sic] were refocused in another Communication published in 2007 (European Commission 2007c), in which the Commission stated: 'Priority should be attached to those infringements which present the greatest risks, widespread impact for citizens and businesses and the most persistent infringements confirmed by the Court.' Apart from cases of non-communication of national measures and non-compliance with Court judgments, the Commission intends to prioritize 'breaches of Community law … raising issues of principle or having particularly far-reaching negative impact for citizens, such as those concerning the application of Treaty principles and main elements of framework regulations and directives.' The 2007 Communication announced that this general guidance would be further specified for each sector, and this is what the 2008 Communication did for the environment sector.

This Communication tends to downplay the importance of enforcement action through infringement proceedings before the Court of Justice and instead stresses a combination of measures aimed at prevention of breaches. The Commission intends to cooperate closely with Member States to help them implement EU environmental legislation and 'solve problems highlighted by citizens and NGOs' through such measures as guidance documents, regular dialogue and support activities. On a trial basis, the Commission will post environmental experts in its Representations in Madrid, Lisbon, Rome and Warsaw 'to help national officials as well as citizens' in implementing environmental legislation.

Where the preventive approach has failed, the Commission will focus its enforcement activities in a way which it describes as more 'strategic,' by giving priority to addressing those breaches of EU environmental law that it considers to be 'fundamental' or 'systemic.' In selecting those cases it intends to apply the following criteria:

- Non-conformity of key legislation viewed as presenting a significant risk for correct implementation of environmental rules and hence their overall effectiveness
- Systemic breaches of environmental quality or other environmental protection requirements presenting serious adverse consequences or risks for human health and wellbeing or for aspects of nature that have high ecological value
- Breaches of core, strategic obligations on which fulfilment of other obligations depends and
- Breaches concerning big infrastructure projects or interventions involving EU funding or significant adverse impacts.

The Communication also stresses the importance of enforcement through national courts in the Member States and, in this respect, the Commission refers to its 2003 proposal for a directive on access to justice in environmental matters. It reiterates its 'view that Community environmental law would be better and more consistently enforced if the proposed directive were adopted.'

Examining the trajectory of EU environmental legislation, we observe that the European Commission has ongoing concerns that the effectiveness of EU directives is undermined because of failures in transposition, implementation and enforcement at the Member State level. The

European Union is able to pass environmental directives without unanimous consent and bring proceedings against Member States for failures to transpose, which may result in financial penalties being applied. However, on several occasions, efforts by the Commission to insert EU institutions more directly into implementation have been rebuffed. There is no agency to monitor implementation or enforcement. Even proposals for EU level capacity to audit Member State activities have been rejected.

These measures concern centralized enforcement. The EU has also sought to promote decentralized enforcement, by encouraging greater citizen access to environmental justice, particularly at the level of member states. However the EU has also highly constrained such activities, by imposing preconditions that limit the impact of such mechanisms for making governance less top-down and more directly responsive to citizens. Thus, while the EU has more powers and greater impact than any international environmental treaty body, Member States have had some success in blocking greater accountability from above and also from below.

NOTES

1. A rather superficial study of the implementation of the Birds Directive, the Seveso Directive and the Bathing Water Directive in all the Member States.
2. A comprehensive study of the implementation of European Commission environmental law in the United Kingdom.
3. Haigh's (1986) is a comparative study of the implementation of the directives on water pollution and waste management in four Member States. Bennett's (1991) is a comparative study of the implementation of the directives on air pollution in the then 12 Member States).
4. See, e.g. Schermers and Waelbroeck (1992, 278): 'The Communities enforce their rules more effectively than other international organizations, in particular as result of the greater homogeneity of their Members and their stronger institutional structure.'
5. 'Analysis of the conditions for its effectiveness make cooperation between administrative implementation research, European comparative law and legislative theory proper seem appropriate' (Schwarze et al. 1994, 91).
6. European Union environmental legislation largely relies on two instruments: directives and regulations. Directives, which are far more common, must be transposed into Member State legislation to take effect. Eujropean Union Regulations, by contrast, apply directly in all Member States.
7. This confusion has been noted by L. Krämer (1992, 209, n. 1), who, observing that 'the notion of "enforcement"... does not seem to be subject to a precise definition which would distinguish it from "implementation,"' has chosen to use the term 'enforcement' as a synonym for 'practical implementation of legislation at ... local, regional or national level,' i.e. application. This option, it seems to me, rather adds to the confusion than reduces it, as the semantic difference between enforcement and

implementation/application is more obvious than that between implementation, transposition and practical application.

8. A good example of the use of this technique in the field of environmental policy is the informal Community network for the implementation and enforcement of environmental law (known as the 'IMPEL' network) in which DGXI cooperates with the environmental inspectorates and other enforcement authorities of the Member States. See: European Commission, 1996a, 18–20.
9. Cf. Krislov et al. (1986, 60): suggesting that 'both conceptually and pragmatically, implementation of and compliance with Community law are issues which must be analysed in the context of decision-making; *in other words, that the process of decision-making does not end with the formal emergence of a Community norm*' (emphasis added).
10. Cf. Krislov et al. (1986, 87): referring to the 'diffuse and fragmentary characteristics' of decentralized enforcement. The Commission itself recently recognized that, even in the field of internal market legislation, judicial enforcement of individual rights is insufficient to ensure the full effectiveness of Community law (European Commission, 1996b, 25).
11. Though it should be noted here that the Court of Justice has held that certain environmental directives may, indirectly, confer rights on individuals (Case C-361/88).
12. Court of Auditors, Special Report No. 3/92 concerning the environment, 18 June 1992, OJ No C 245, 23.9.1992, p. 1, at p. 19.
13. Though it is somewhat puzzling that these measures, which are in fact no more than existing obligations of the Member States resulting directly from the Treaty and the relevant secondary legislation itself and can therefore hardly be considered as revolutionary, are listed under the heading 'practical *reforms* to be undertaken' (European Commission, 1992, 76 (emphasis added)).
14. For a general overview, see Somsen, 1996.
15. According to Krämer (1995, 141), reports submitted by Member States pursuant to reporting requirements in directives are 'rarely a source of information on effective implementation of Community environmental rules.'
16. Whether it actually 'leaves little room for strategic decision-making' on enforcement priorities by the Commission, as Macrory (1992, 363) has claimed, is questionable, in view of the discretionary nature of the Commission's recourse to Article 169 proceedings and the ample evidence that its enforcement policy is strongly influenced by political expediency; see Williams, 1994.
17. For an analysis of this debate, which became a *cause célèbre* during the political struggle between the European Parliament and the Council over the establishment and mandate of the European Environment Agency, see Collins and Earnshaw 1983, 238–42.
18. Winter (1996, 206–8) quite pertinently rejects this approach on the grounds that entrusting the enforcement of EC environmental law in individual cases entirely to national procedures would stunt the further development of the law and confine the Court of Justice to the 'anaemic exercise' of reviewing formal transposition only.
19. See OJ No C 242, 21.8.1996, p. 6.
20. The Member States concerned are Germany and Italy, who have failed to comply with Court judgments concerning the application of several environmental directives. See European Commission, 1997b. Since then, similar actions have been brought against Greece (in a case involving the implementation of two directives on waste management) and Belgium (in a case concerning the implementation of the Birds Directive). See European Commission, 1997c; 1997d.
21. For a similar analysis focusing on the Common Agricultural Policy, see also Bronkhorst 1993.

22. Both the Communication and the Action Programme address improvements in the legislative process under the general heading of 'implementation.' While the drafting and negotiation of Community environmental legislation obviously has an important impact on its eventual *effectiveness*, it seems questionable from a conceptual viewpoint to regard it as part of the actual process of *implementation*.

REFERENCES

Bennett, G. (ed.) (1991), *Air Pollution Control in the European Community: Implementation of the EC Directives in the Twelve Member States*, London: Graham and Trotman.
Bronkhorst, H.J. (1993), *Deugdelijkheid van EG-regelgeving*, Deventer: Kluwer.
Case 26/62, *Van Gend en Loos*, [1963] ECR 10.
Case 65/76, *Derycke*, [1977] ECR 35.
Case C-213/89, *Factortame*, [1990] ECR I-2433.
Case C-361/88, *Commission v. Germany*, [1991] ECR I-2567.
Collins, K. and D. Earnshaw (1993), 'The Implementation and Enforcement of European Community Environment Legislation,' in D. Judge (ed.), *A Green Dimension for the European Community: Political Issues and Processes*, London: Frank Cass, pp. 213–49.
Court of Auditors (1992), Special Report No. 3/92 concerning the environment, 18 June 1992, OJ No C 245, 23.9.1992.
Curtin, D. (1990), 'Directives: the Effectiveness of Judicial Protection of Individual Rights,' *Common Market Law Review*, **27**, 709–39.
Ehlermann, C.D. (1988), 'Opening Speech at the IVth Erenstein Colloquium,' in H. Siedentopf and J. Ziller (eds), *Making European Policies Work: The Implementation of Community Legislation in the Member States*. London: SAGE Publications, pp. 143–50.
European Commission (1985), EC Doc. COM (85) 310 final, 14 June 1985.
European Commission (1986), EC Doc. COM (86) 485 final, 9 October 1986.
European Commission (1992), EC Doc. COM (92) 23 final, 27 March 1992, Vol. II.
European Commission (1996a), 'Implementing Community Environmental Law,' Communication from the Commission, EC Doc. COM (96) 500 final, 22 October 1996, Annex I.
European Commission (1996b), EC Doc. COM (96) 520 final, 30 October 1996.
European Commission (1997a), Press Release IP/97/5, 8 January 1997.
European Commission (1997b) Press Release IP/97/63, 29 January 1997.
European Commission (1997c), Press Release IP/97/571, 26 June 1997.
European Commission (1997d), Press Release IP/97/1115, 15 December 1997.
European Commission (2002a), EC Doc. COM (2002) 141 final.
European Commission (2002b), EC Doc. COM (2002) 725 final.
European Commission (2002c), EC Doc. SEC (2002) 1041.
European Commission (2003a), EC Doc. COM (2003) 667 final.
European Commission (2003b), EC Doc. COM (2003) 624 final.
European Commission (2003c), EC Doc. COM (2003) 669 final.
European Commission (2005), EC Doc. COM (2005) 570 final.

European Commission (2006), EC Doc. SEC (2006) 1143.
European Commission (2007a), EC Doc. COM (2007) 707 final.
European Commission (2007b), EC Doc. COM (2007) 51 final.
European Commission (2007c), EC Doc. COM (2007) 502 final.
European Commission (2008a), European Commission Directive 2008/99/EC on the protection of the environment through criminal law.
European Commission (2008b), Communication from the Commission on implementing European Community Environmental Law, EC Doc. COM (2008) 773 final.
European Council (1990), Statement of the European Council, Dublin, 26 June 1990, *Bull. EC*, 1990/6.
European Council (1993), Council Resolution of 8 June 1993 on the quality of drafting of Community legislation, OJ No C 166, 17.6.1993.
European Council (1997), Council Resolution of 7 October 1997 on the drafting, implementation and enforcement of Community environmental law, OJ No C 321, 22.10.1997.
European Parliament (1984), EP Resolution of 11 April 1984 on the treatment of waste in the European Community, OJ No C 127, 14.5.1984.
European Parliament (2008), Resolution of 20 November 2008 on the review of Recommendation 2001/331/EC providing for minimum criteria for environmental inspections in the Member States. From: http://www.europarl.europa.eu/sides/getDoc.do?pubRef=-//EP//TEXT+TA+P6-TA-2008-0568+0+DOC+XML+V0//EN&language=EN
Final Act of the Intergovernmental Conference (1992), 'Declaration on the Implementation of Community Law,' in *Treaty on European Union*, Office of Official Publications of the European Communities, Luxembourg.
Glim, C.M.S (1990), *European Environmental Legislation – What Does it Really Mean?* Delft: Eburon.
Haigh, N. (1986), *Comparative Report: Water and Waste in Four Countries – A Study of the Implementation of the EEC Directives in France, Germany, Netherlands and the United Kingdom, European Community Environmental Policy in Practice*, Vol. 1, London: Graham and Trotman.
'How to Strengthen the Effectiveness of Community Law' (1991), *Common Market Law Review* **28**, 711–16.
Jadot, B. (1987), 'Le justiciable et l'inexécution, en droit belge, des directives européennes en matière d'environnement,' *Aménagement-Environnement*, pp. 34–42.
Jadot, B. (1989), 'L'effet direct des directives européennes en matière de protection de l'environnement,' in L. Lavrysen (ed.), *Milieurecht – Recente ontwikkelingen*, Vol. 2, Brussels: E. Story-Scientia, p. 241.
Jans, J.H. (1990), 'De kwaliteit van de Europese milieuregelgeving,' *Sociaal-Economische Wetgeving*, **38,** 503–26.
Joined Cases C-6/90 and C-9/90, *Francovich*, [1991] ECR I-5357.
Judgment of 13 September 2005, Case C-176/03, *Commission v Council.*
Krämer, L. (1988), 'Du contrôle de l'application des directives communautaires en matière d'environnement,' *Revue du Marché commun*, pp. 22–28.
Krämer, L (1992), *Focus on European Environmental Law*, London: Sweet and Maxwell.

Krämer, L. (1995), *E.C. Treaty and Environmental Law*, 2nd ed., London: Sweet and Maxwell.

Krämer, L. (1996), 'Direct Effect of EC Environmental Law,' in H. Somsen (ed.), *Protecting the European Environment: Enforcing EC Environmental Law*. London: Blackstone Press, pp. 99–134.

Krislov, S., C.D. Ehlermann and J. Weiler (1986), 'The Political Organs and the Decision-Making Process in the United States and the European Community,' in M. Cappelletti, M. Seccombe and J. Weiler (eds), *Political Organs, Integration Techniques and Judicial Process, Integration Through Law*, Vol. 1, Book 2, Berlin: Walter de Gruyter, pp. 3–110.

Macrory, R. (1992), 'The Enforcement of Community Environmental Laws: Some Critical Issues,' *Common Market Law Review* **29**, 347–69.

Macrory, R. (1996), 'Community Supervision in the Field of the Environment,' in H. Somsen (ed.), *Protecting the European Environment: Enforcing EC Environmental Law*, London: Blackstone Press, pp. 9–22.

Pagh, P. (1995), 'The 'Direct Effect Doctrine' in EC Environmental Law,' *Nordic Journal of International Law*, **64**, 23–57.

Plaza Martin, C. (1994), 'Furthering the Effectiveness of EC Directives and the Judicial Protection of Individual Rights Thereunder,' *International and Comparative Law Quarterly*, **43**, 26–54.

Prechal, S. (1995), *Directives in European Community Law*, Oxford: Clarendon Press.

Schermers, H.G. and D. Waelbroeck (1992), *Judicial Protection in the European Communities*, 5th ed., Deventer: Kluwer Law and Taxation.

Schwarze, J. U. Becker and C. Pollak (1994), *The Implementation of Community Law: Studies in the Legislative and Administrative Policies of the European Community and its Member States*, Baden-Baden: Nomos Verlagsgesellschaft.

Siedentopf, H and J. Ziller (eds) (1988), *Making European Policies Work: The Implementation of Community Legislation in the Member States*, London: SAGE Publications.

Snyder, F. (1993), 'The Effectiveness of European Community Law: Institutions, Processes, Tools and Techniques,' *Modern Law Review* **56**, 19–54.

Somsen, H. (ed.) (1996), *Protecting the European Environment: Enforcing EC Environmental Law*, London: Blackstone Press.

Wägenbaur, R. (1991), 'The European Community's Policy on Implementation of Environmental Directives,' *Fordham International Law Journal*, **14**, 455–77.

Weatherill, S. (1995), *Law and Integration in the European Union*. Oxford: Clarendon Press.

Weiler, J. (1988), 'The White Paper and the Application of Community Law,' in R. Bieber, R. Dehousse, J. Pinder and J. Weiler (eds), *1992: One European Market? A Critical Analysis of the Commission's Internal Market Strategy*, Baden-Baden; Nomos Verlagsgesellschaft, pp. 337–58.

Williams, R. (1994), 'The European Commission and the Enforcement of Environmental Law: an Invidious Position,' *Yearbook of European Law*, **14**, 351–99.

Winter, G. (1996), 'On the Effectiveness of the EC Administration: The Case of Environmental Protection,' *Common Market Law Review*, **33** 689–717.

11. What is multilevel environmental governance? When does it work?

Inger Weibust

Having examined the experiences of several multilevel systems, what common patterns emerge? Do we observe weakness in central governments and fading of authority? Is there significant autonomous action at the local level? What are the findings with regard to the policy efficacy of different variants of multilevel governance? This chapter begins with a summary of Daniel Fiorino's survey of the literature on the conditions of environmental policy effectiveness. This is followed by an analysis of the book's empirical chapters. The second section contrasts the multilevel governance of climate policy and water. The third section contrasts the empirical findings with Hooghe and Marks' original typology (2001) of multilevel governance (MLG). In light of the shortcomings of Hooghe and Marks' MLG typology, the fourth section discusses the options for categorizing the governance patterns just presented. The final section assesses the performance of various multilevel governance arrangements, highlighting the importance of the shadow of hierarchy.

Daniel Fiorino's chapter surveys the literature on explaining cross-national variation in environmental policy performance, particularly studies examining the impact of national income, democracy and aspects of political system (e.g. corporatist, unitary, proportional representation). Noting the presence of multiple possible causes of policy performance across cases, Fiorino concludes that it is not possible to control for those other factors to draw absolute conclusions about the impact of governance structures. He cites Walti, who points to a unique positive synergy for environmental performance in those countries which are high income, corporatist and strongly federalist. Fiorino offers illustrations of possible impacts of multilevel governance structures at different stages of the policy process (e.g. agenda setting, implementation, etc.). He closes by advocating further research into the impact of these interaction effects, as well as identifying the optimal level of jurisdiction by type of environmental problem.

EMPIRICAL EVIDENCE ON MULTILEVEL GOVERNANCE FROM THE CASES

Because the Multilevel Governance concept was developed to describe the European Union (EU) (Hooghe and Marks 2001), it is appropriate to begin with Marc Pallemaerts' description of pathways for enforcement of EU environmental legislation. His analysis offers insight into complex patterns of authority and how they have shifted over time. Under the European Union's system of *administrative federalism*, Member States have some autonomy in implementation, as in the German federal system (Peters 2006, 179; Schneider 2006). The extent of that autonomy is subject to ongoing debate between the European Commission and Member States, as represented on the European Council. Only a few EU environmental directives have direct effect: the vast majority have to be enacted by Member State legislation. However the European Commission does compel correct transposition of its directives into Member State legislation. Incorrect transposition can result in a Member State being brought before the European Court of Justice and, since 1993, financial penalties can be applied.

Pallemaerts describes two pathways for enforcement of EU environmental directives. The first are infringement proceedings, brought by the European Commission, usually as a result of investigations of Member State transposition by the European Commission. The second are citizen complaints. Thus, there is top-down enforcement of EU directives, by the Commission against Member States. But through the citizen's complaint procedure, there is also some bottom-up enforcement of transposition into Member State policies. However the impact of the bottom-up mechanism is blunted by the Commission's substantial involvement in the process. The Commission chooses which and how many citizen complaints to bring forward. This form of citizen engagement is far weaker than that seen in the United States (US), where citizens or groups can go directly to court to seek the enforcement of a large number of federal environmental statutes (Heyes 2000, 120).

Implementation and EU oversight of transposition have long been sources of contention between the European Commission and Member States. The European Commission has pushed for greater powers, on the grounds of ensuring policy effectiveness and guaranteeing citizens' rights. At times the Commission's rhetoric suggests it favours greater decentralization, although sometimes these pronouncements seem contradictory. Pallemaerts notes, for example, that in 1996 the Commission announced that it would ease Commission control, while also seeking better enforcement of EU directives.

More recently, the Commission has asserted its authority by pushing for greater consistency in inspections, the aspect of implementation that has, historically, been considered to lie most clearly within exclusive Member State competences. On grounds of consistency and enforcement, the Commission had proposed an EU-wide inspectorate to assist or audit national inspections, but this proposal was firmly rebuffed by Member States. Nevertheless, in 2008 the Commission passed a directive which sets more uniform requirements for criminal penalties in Member State environmental legislation. A push for greater uniformity in penalties would seem to encroach on Member State discretion in implementation.

In environmental governance the European Union retains many hierarchical elements, particularly in Commission oversight of transposition of directives. Pallemaerts notes that the Commission has become somewhat less zealous in enforcement, preferring prevention to punishment, in part because of increased workload due to the accession of new Member States. He notes that while negotiation is preferred, the Commission retains the credible threat of litigation as a bargaining chip. In general, the Commission does not otherwise seem keen to expand Member State autonomy, continuing to push for greater uniformity in policy implementation. The Commission may supplement the stick with a carrot, but its goal in ensuring effective, relatively uniform, environmental legislation seems unwavering.

Similarly the Commission shows little inclination to 'give more power to the people.' While the citizen's complaint procedure permits public participation, allowing citizens to do an end run on their national governments, the Commission retains tight control of the process. Thus citizens and groups that feel shut out by the policy process at the national level do not have guaranteed access at the EU level. The Commission has placed conditions on which complaints will be allowed to go forward, further limiting the scope of this tool. Pallemaerts notes that the Commission is bringing fewer complaints forward and suggests that the new criteria are sufficiently restrictive that they significantly reduce citizen access.

In summary, in Pallemaerts' description of enforcement of EU environmental policy, we find Member States whose autonomy has been constrained by EU directives. There is continuing contention between Member States and the European Commission over remaining areas of exclusive Member State competencies, particularly in the area of inspections and Member State enforcement of environmental regulations. While Member States have lost some autonomy, the governance arrangements retain strong hierarchical elements, particularly between the EU and Member States.

In their chapter, David Benson and Andrew Jordan trace how the European Union has grappled with the question of governance scale (in other words, 'optimal jurisdiction') for water quality policy. Since 1987 these debates have revolved around the interpretation and implementation of subsidiarity. They note that although many advocates of subsidiarity saw it as a needed brake on the centralizing tendencies of the European Commission, in application, subsidiarity has contributed to *greater* EU involvement in the policy area. Thus, while understood by many to mean devolving decision making to the lowest possible level and slowing the incursion of EU authority, for European Commission President Jacques Delors, the principle justified greater EU intervention on grounds of the common good. The malleability of the concept was key to securing its broad support.

This malleability is possible because subsidiarity does not simply dictate that policy be made at the lowest level, but at the lowest effective or practical level, subject to a number of tests. In this issue-area, these tests have been interpreted to mean that the European Union is the lowest possible level, given the transboundary impacts of European water pollution. Thus, while the Water Framework Directive is organized around the scale of river basins, it is the EU that dictates that this must be the case. Member State adoption of this form of governance is mandatory. Thus, whatever the merits of maintaining national control, the objective of using subsidiarity to reassert national autonomy in water quality must be judged a failure.

FINDINGS FROM MULTILEVEL GOVERNANCE OF CLIMATE POLICY

In examining the local element in EU climate policy, Kristine Kern describes a situation which is not reducible to top-down or bottom-up narratives. Even where the mode of interaction is primarily vertical (not horizontal, between cities), there is complex interplay between local governments and central governments or EU institutions. In some EU Member States, action by local governments is supported by subsidies or capacity-building programs funded by the national or subnational government. These appeared first in unitary Member States with relatively autonomous local authorities. Because in Germany, the regional *Länder* governments have a more direct relationship with local government than the German federal government does, the German federal government's efforts to engage local government have tended more to voluntary actions, such as sponsoring a contest for German cities excelling in climate policy.

As predicted by MLG theory, cities have been building direct institutional links with the European Union, bypassing their Member State governments (Hooghe and Marks 2001). However the European Union has provided financial incentives for these activities. As Kern puts it, 'EU funding has become an important driver for bottom-up Europeanization.' Thus these local government-based activities cannot be seen as straight forward examples of 'bottom-up' initiatives to engage the European Union level of governance.

Although Kern focuses on the process of Europeanization in climate policy at the local government level, she also offers a counter example to MLG theory's account of the weakening of national authority. Kern finds that, in this policy sphere, the United Kingdom behaves like an archetypal unitary state: it mandates climate policy targets for local councils and monitors their performance of these targets. This example of a top-down approach, which would be unacceptably intrusive in many democracies, suggests the predictions of MLG have not yet come to pass everywhere, not even in the EU, which is the paradigmatic example for MLG theory.

Kern notes that such mandates are not found in other Member States such as Germany, the Netherlands or in the Nordic countries where local authorities enjoy greater autonomy from national governments. However, local government autonomy from national governments is not a new development in these countries, thus this observation does not offer strong support for claims of MLG as an inevitable *process* of weakening of the national state. More observations over time would be required to support the claim that a progressive weakening of national authority vis-à-vis lower level governments is occurring.

What conclusions can we draw about the efficacy of local government climate change policy? Kern notes that many cities and city networks for climate change have set targets for greenhouse gas (GHG) emissions reduction that were more ambitious than those set by EU Member States. It is unclear, however, how many of these ambitious targets have been reached. As we know from the example of the Kyoto Protocol, the ambitiousness of a target is no guarantee of achievement! For example, it seems unlikely that most cities in the Climate Alliance (a network of cities within the EU) met the Alliance's initial goal of a 50 per cent cut in

CO_2 emissions (on a 1990 baseline) by 2010. Kern also notes that, while pioneering cities are active participants in European climate networks, there is no guarantee that best practices are diffusing to laggards. With regard to the prospects for diffusion of best practices, Kern notes that differences between cities persist even within the same country. For real reductions in GHG emissions, it matters what the average city does, not just the pioneers.

Turning to American climate policy, Barry Rabe's account begins with a description of 'state dominated' American climate policy, due to Presidents and/or a Senate that were hostile to greenhouse gas reductions. When state governments stepped into the breach, it was a classic example of compensatory federalism, where one level of government makes up for the deficiencies of the other. This scenario also accords with MLG theory's predictions about weakening of central governments. However, whereas MLG theory treats this as a universal, secular trend, Rabe's analysis suggests that, at least for the American case, the trend will not necessarily continue. It appears that, in his second term, President Obama intends to put climate change mitigation back on the policy agenda.

American climate policy at the state level has followed the classic pattern of policy innovation and diffusion described in the federalism literature. Furthermore, not only are these policies diffusing across states, they are also becoming more rigorous over time (Rabe 2009, 73). However this state-level activism does not imply that state governments have abandoned self-interest. Some of these policies have been introduced under the rationale of stimulating state economic development, particularly those promoting alternative sources of energy. Rabe also notes that states have downplayed potential costs for citizens and have been averse to taxes which might increase energy costs. Nonetheless, a few states have blocked the construction of new coal fired power plants which, indirectly, may lead to higher electricity costs. Some of the state policy options do impose costs, but these are typically *outside* the state's jurisdiction, e.g. California's revised vehicle emissions standard. Rabe also observes that there are rewards to politicians and officials who act as policy entrepreneurs. Entrepreneurship can yield concrete benefits in terms of increased visibility, which can boost a political career or result in national prominence. Thus leadership and self-interested behaviour can go hand in hand.

Although state innovation is consistent with a 'bottom-up' narrative and a picture of a weakening national level of government, it is not the whole story here. Although some state environmental officials have consistently argued for more state autonomy and a more circumscribed federal role, other state leaders have been pushing the federal government to act. In 2003, 13 state attorneys general took the US federal government to court to force it to regulate greenhouse gas emissions under the federal Clean Air Act. Here we have a paradox of subnational governments using their autonomous powers to try to force the national government to *occupy*, not *vacate*, a policy space. It is also noteworthy that state governments chose to advance this agenda through the courts, not simply advocacy or negotiation. They chose a forum which would yield an authoritative outcome. This type of state level initiative makes sense in light of the competitiveness and leakage issues that unilateral state action on climate change cannot address.

Furthermore Rabe argues that state level innovation must be seen in an intergovernmental context, not in isolation. Elsewhere Rabe has attributed the growth in state level climate policy capacity to federal grants and technical assistance, beginning with the 1990 State and Local Climate Change Program (Rabe 2007, 435). There is also the horizontal level of interstate cooperation. Organizations representing all or most of the states are lobbying for federal measures that protect existing state policies, as well as seeking financial support from the federal government for their policies. Some states may be breaking away from the pack, in an attempt to secure federal policies more favourable to their own unique circumstances. In particular the states that have been climate policy pioneers are eager to get credit for reductions already undertaken. These are examples of political choices made in the shadow of hierarchy. Under the Bush administrations states showed leadership on the climate file, while the federal government was hostile or, at best, inactive. However while the federal government was not active in this period, the possibility remained that it would assume a more active role in the future and this may have conditioned state government behaviour.

Rabe presents several scenarios for future US federal climate action, which differ in how coercive or cooperative the federal-state relationship is. Under the most coercive option, *total pre-emption*, the federal government imposes regulations, which override existing state measures, creating a uniform national standard. Total federal pre-emption has occurred in a range of policy areas and could be applied to portions of US climate action, such as the introduction of a national cap and trade scheme. Such a non-negotiated assertion of federal authority (and uniformity) would be entirely at odds with the predictions of MLG theory.

At the opposite end of the continuum is *collaborative federalism*, whereby the federal government would establish an overarching framework specifying required emissions levels, but leave it to the states to decide how these targets were to be achieved. The federal government would also retain the capacity to sanction states for failure to achieve these targets. This scenario is clearly far less centralizing than total pre-emption, but it retains key hierarchical elements. An arrangement whereby the US federal government can unilaterally set overall emissions limits and demand compliance with them is not characteristic of a disempowered federal government and non-hierarchical governance.

David Gordon and Douglas Macdonald's chapter seeks to explain the variation in national climate policy outcomes between Australia and Canada. Both countries showed ambivalence towards the Kyoto process and their targets from it, but at the national level, Australia eventually passed a carbon tax. In contrast, as of mid-2013, Canada had no binding national policies to cut GHG emissions. These two countries share many similarities: federal systems; resource dependent export oriented economies; and high per capita GHG emissions. As federations, both have relatively autonomous sub-national governments. Gordon and Macdonald argue that the greater strength of institutions for intergovernmental coordination help explain success in the Australian case by making subnational governments more receptive to national action.

While the Australian outcome is not reducible to simple federal pre-emption, it is worth noting it was also not produced by policy diffusion across subnational governments nor by horizontal cooperation, absent federal involvement. Australia's national carbon tax was approved through its federal bicameral legislature, despite the opposition of some states. Although states are represented in Australia's Senate, passage of the legislation did not require unanimous approval by all states and territories. Citing Braun (2006) on Australian economic policy coordination, Gordon and Macdonald note that federal leadership may be uniquely important in policy coordination, because of the federal government's capacity to use political and fiscal levers in order to create larger win-sets for interjurisdictional agreement.

In the final chapter on climate policy, Douglas Macdonald presents three cases of climate burden sharing within federal systems (Germany, the EU and Canada) and seeks to explain the successful allocation of emissions cuts in the EU and Germany as against the failure to allocate national reductions across Canadian provinces. In the case of the EU, the proposed allocation of cuts across member states was explicit. For both Germany and Canada, the differential regional impact of emissions reduction plans remained implicit, with no explicit subnational targets assigned. In the Canadian case, participants in the process felt explicit allocation would be too divisive. However, the implicit approach did not prevent conflict: Canadian federal efforts at crafting a national position aroused strong opposition from Alberta and Saskatchewan, whose per capita emissions have been approximately seven times those of the lowest ranked province. Any compensation offered to these two provinces would have been very controversial in other provinces, particularly Quebec, which is a relatively low per capita emitter.

Macdonald points to a confluence of motivation (preferences) and strategies, but primarily to institutions to explain the outcome. Federal institutions which rely upon unanimous decision making create opportunities for lower level governments to block national (or EU level) initiatives. Furthermore, less routinized federal institutions create the possibility of exit, by either the federal or subnational governments. In Canada's highly decentralized federation, Alberta had the opportunity to veto or exit federal-provincial negotiations on GHG reductions.

While the European Union, like Canada, relied on consensus among Member States, German and European federal institutions did not offer the same opportunity for exit. German institutions offer little opportunity for a veto by an individual Land. At the federal level, the *Länder* governments are represented directly in the *Bundesrat* whose members they appoint. The *Bundesrat* passes legislation with a simple majority. Furthermore, once federal legislation is passed by the *Bundestag* and *Bundesrat*, it is binding upon all *Länder*, eliminating the possibility of exit during implementation. Similarly, although EU Member States lobby to influence EU legislation and can vote against it in the European Council, if a directive passes, individual Member States cannot refuse to participate, without incurring a penalty.

But if consensus decision making makes it easier to block regional allocations, why was the EU successful in assigning explicit Member State targets while Canada failed to achieve any national plan for cutting emissions? Macdonald argues that differences in motivation and strategy matter here. Specifically, the European Commission was much more motivated to reach agreement, in part because it sought to play a leadership role internationally. By contrast, no Canadian federal government, Liberal or Conservative, had expressed interest in being an international climate leader. With regard to questions of strategy, Macdonald suggests that the capacity to offer side payments can be significant. The European Union's decision to have explicit national emissions reduction allocations made discussion of compensation possible.

An alternative interpretation is that the differences between outcomes in three systems is somewhat overstated. Germany and the European Union may be only marginally more successful at making tough choices than Canada. The arbitrary choice of a 1990 baseline for emissions reductions under the Kyoto Protocol was uniquely favourable to Germany, because the German baseline was calculated to include East

Germany's highly polluting industries, which mostly ceased production upon reunification. Thus Germany was able to achieve significant emissions reductions without taking any further action. Together, Germany and the United Kingdom's reductions accounted for more than 100 per cent of the total EU Kyoto reduction, which meant that emissions elsewhere in the EU had room to grow, substantially in some cases (Harrison and Sundstrom 2007, 4). These factors do not count for the entirety of cuts made within the EU: actual emissions reductions did occur.

In contrast Canada did not obtain any similar advantages. Kathryn Harrison (2007, 92) notes that, for Canada, meeting its Kyoto target would have required a 29 per cent reduction from the 'business as usual' scenario for that time period. Compared to Germany and the EU, Canada faced challenges as a major oil producer and also because its population and economy have continued to grow (Simpson et al. 2007, 83). Canada's highly integrated economic relationship with the US, which did not ratify Kyoto, also inhibited action by Canadian governments.

Thus, although the emissions reduction targets were similar, the *effort* required to achieve them was not. Similarly, although the European Union was able to allocate permits for its Emissions Trading Scheme (ETS), the price fell almost to zero. In the first phase (2005–2007), the ETS did not require emitters to reduce emissions, thus tough decisions about allocation were not made until Phase II, beginning in 2008, which cut the total supply of permits by 6.5 per cent (European Commission 2013). However, in the wake of the 2008 financial crisis and economic downturn, there was an excess supply of permits. The price of permits has remained low and will face downward pressure given legislated mandates automatic increases in the total number of permits. Time will tell if the European Union and Germany are able to make tough choices on emissions cuts and stick to them.

FINDINGS FROM MULTILEVEL GOVERNANCE OF WATER

Timothy Heinmiller's chapter comparing governance of irrigation presents a variety of distinctive multilevel arrangements, with consequences for policy outcomes. In Canada, the Alberta provincial government has had and continues to have, very extensive autonomy in this policy area and is only minimally constrained by federal prerogatives. Australian states used to have a similar level of autonomy until interstate water allocation conflicts gave rise to a stronger role for the Australian federal government and interstate governance. State-level irrigation policy making now occurs in a nested context in relationship to the Australian federal government and federal-state intergovernmental bodies for water governance. In the United States, both state and federal governments play major roles in water allocation decisions. However their pattern of interaction is closer to being in a parallel, rather than the nested relationship that is found in Australia.

In Alberta, all irrigation politics takes place at the provincial level. If an interest group can win at the level of the provincial government, it largely dictates the policy. Thus far in Alberta, the irrigators have had far more clout at the ballot box than environmentalists in the rural dominated provincial legislature. Successive irrigation reforms in Alberta have favoured irrigators over the interests of environmentalists or urban water users.

Because institutions in Australia are nested, irrigator-environmentalist conflicts have played out on several levels. The federal government and state governments play a role, but intergovernmental bodies have also been important. While federal and intergovernmental venues have been most sympathetic to environmentalists' preferences, state venues have been most sympathetic to irrigators' preferences. Irrigators have had to yield substantial ground to the conservationists as Australian water policy has evolved.

In the California case, the US federal government has for decades possessed significant legal authority in this domain, but had generally chosen to exercise those powers sparingly until the 1970s. Prior to that time the California government was relatively autonomous in making choices about irrigation. The existence of two parallel fora, with some federal supremacy, created opportunities for venue shopping on the part of environmentalists. When environmentalists' prospects looked unfavourable at the state level, they were able to shift their efforts to the federal level.

These differences in governance seem to have had consequences for policy outcomes. Australia and the US have been more successful in protecting instream (non-consumptive) flows of water for environmental benefits. In contrast, Alberta has environmental protection as one policy objective among many, but consumptive uses by irrigators take priority under its First in Time, First in Right (FITFIR) legal regime. The variation in these outcomes occurred because the federal governments of Australia and the US have played a stronger role in those systems, weakening the capture of water policy by irrigators which has tended to occur at the state and provincial level. Although California also has a

FITFIR legal regimes for water rights, federal legislation and the courts have forced the state to consider downstream riparians, as well as allocations for instream flow, for ecological benefits.

Heinmiller's findings are consistent with observations from the US literature showing that lower levels of government are more likely to favour development and business interests and that environmentalists have more influence at the federal level, rather than at the state or local level (Koontz 2002; Layzer 2008). It is intriguing that this pattern holds in all three federations, not just the American case where this literature was developed.

Irrigation also features in Inger Weibust's chapter on allocating water between jurisdictions, because irrigation is typically the largest consumptive use of a river. The chapter assesses the governance structure of river allocation regimes in Canada, the US and the EU, as well as the policy performance of these regimes.

In Canada, the most important interprovincial rivers all involve the province of Alberta as the upstream riparian and of these, only the South Saskatchewan River is covered by an active interprovincial agreement. Canadian governments are proud of the low key, consensual manner in which this agreement has operated for decades. In theory, the federal government could be called upon to settle an interprovincial dispute but this has never occurred. It is unclear, however, whether this consensual arrangement will be equal to the task of ensuring water for downstream provinces as urban demand grows and climate change diminishes flow. Downstream provinces have no guaranteed right or entitlement to water on any of these interprovincial rivers. The users with the strongest legal entitlement to the South Saskatchewan River are Alberta irrigators. The agreement already shows deficiencies in failing to guarantee minimum flows for ecological purposes.

The American system is almost a polar opposite of the Canadian regime: it is legalistic, rigid and often highly contentious and very slow. Although state governments play an important role, their actions take place against a backdrop of several federal laws, as well as the authority of the US Supreme Court to enforce interstate agreements and enforce federal laws.[1] Thus there is a strong element of shadow of hierarchy, if not actual hierarchy, in the system.

The shadow of hierarchy which federal legislation and the Supreme Court represent introduces both certainty and uncertainty into the governance regime. There is certainty, in that the US Supreme Court guarantees state that water apportionment agreements they negotiate among themselves will be enforced, although probably not quickly. There is substantial uncertainty as well, as the state of Georgia discovered in 2009. Turning to the US Supreme

Court can be risky because the many and diverse federal laws involved in water apportionment make it very difficult to predict how the Supreme Court might rule in any given dispute.

Because the Supreme Court established the doctrine of *equitable apportionment* in 1907, the American regime ensures that downstream riparians are not at the mercy of upstream jurisdictions. The Supreme Court has ruled repeatedly that upstream states do not enjoy entitlements to water simply by virtue of being upstream. Over time interstate apportionment agreements have done a better job of recognizing the ecological value of instream uses.

In the European Union, the Water Framework Directive of 2000 provides the backdrop to all water policy in the European Union. Because Member States have until 2015 to fully implement it, it is too soon to pass definitive judgment on its impact. The Directive contains no provisions for allocating rivers between Member States, although there have been ongoing river allocation disputes between some Member States. The Directive hopes to prevent water shortages through better demand management, particularly by requiring metering and pricing of all water uses.

Any disputes over allocation between Member States are to be resolved by reference to international law. However, there is considerable ambiguity in international law, which is vague and even contradictory on these issues. Furthermore the resolution of two current river allocation disputes in the EU has been blocked by subnational governments. Because subnational governments do not enjoy legal personality under international law, it is not clear how international law would be helpful in resolving these disputes. It is possible that the European Union may be pulled in to arbitrate, if not actually settle, these disputes.

CONTRASTING MLG OF CLIMATE CHANGE AND WATER

Are there common patterns across these two domains of climate and water policy? There are important differences between these policy domains. The most important is that climate change mitigation, that is cutting emissions (rather than adaptation to climate change) will always be a global problem, whereas water policy solutions can be local or regional and often do not require concerted international action. The chief similarity is that both policy domains can require making tough policy choices, against a backdrop of hard limits imposed by nature.

The similarities can be seen in allocations of permits for GHG emissions and irrigation. There are strong political incentives to overallocate permits in both cases, although this problem is not unique to multilevel governance. In the case of water, there may be more permits granting rights to water than there is actual water flow in the river. Heinmiller gives examples of rivers in Alberta, Australia and California which are over allocated. Even if a river is not overallocated, as calculated on the basis of average annual flow, a drought may result in permits exceeding the total actual flow, because permits are usually based on a fixed volume of water, not a proportion of actual flow. In those cases, where a cap has been imposed on total water withdrawals, it typically locks in all existing permits, without reducing the total volume to which water users are entitled.

There is an analogous problem in allocating permits for GHG emissions. For an emissions trading scheme to reduce emissions, permits must be at least somewhat scarce: the total quantity of emissions covered by permits must be less than current emissions. A market for permits will only arise if producers face the choice of: 1) reducing their emissions in line with their permit; or 2) purchasing a permit from another emitter who is better able to reduce emissions. One of the reasons for the poor performance of the European Emissions Trading Scheme is that, in the initial Phase I period, many Member State governments were found to have distributed ETS permits in excess of actual emissions (Ellerman and Buchner 2008). Under such a scheme only economic growth over time would make permits scarce whereas an economic downturn exacerbated the situation.

What does MLG mean for these problems? Hierarchy seems beneficial, if not sufficient, for addressing these problems. In the case of water, comparisons are possible between Canada and other federations, as well as the US and Australia over time. Reduced state autonomy, due either to federal courts, intergovernmental institutions or federal legislation, has been consistent with better integration of the environment into water allocation schemes, as well as greater equity between upstream and downstream riparians. Furthermore, in Australia and the US, recourse to a higher level of government has enabled the authoritative resolution of disputes over allocation between subnational jurisdictions and competing uses. (This may not be sufficient to resolve conflicts between users *within* subnational jurisdictions.) That is not to say these resolutions have been achieved quickly or without acrimony. It will be interesting to see how water governance in Canada and the EU evolves as climate change makes water supply a more salient issue in places that have historically had adequate supplies.

Similarly, although the European Emissions Trading Scheme has not yet yielded a robust market for permits, as Macdonald notes in his chapter, the European Union did at least succeed in introducing a form of cap-and-trade, which was EU wide. Although the US has seen regional climate change initiatives, these have not achieved coverage of a majority of states, never mind nation-wide coverage. These regional initiatives have diffused but not very widely. Unlike the American case, there is little evidence of spontaneous diffusion of climate change policies, such as a carbon tax, from Quebec and British Columbia to the remaining Canadian provinces and territories.

DOES ENVIRONMENTAL MLG À LA HOOGHE AND MARKS EXIST?

Although the European Union was the leitmotif and basis for developing MLG theory, trends in environmental governance in the European Union do not fit Hooghe and Marks' empirical description. The evidence here supports Tanja Börzel's (2010, 191) revisionist conclusions about EU governance in general,

> First, the EU relies heavily on hierarchy in the making of its policies. Its supranational institutions allow for the adoption and enforcement of legally binding decisions without the consent of (individual) Member States. Second, network governance, which systematically involves private actors, is hard to find. EU policies are largely formulated and implemented by public actors.

While Member State governments have lost some authority, this does not mean that authority in general is weakening. In environmental policy, the institutions of the European Union, such as the European Commission and the Court of Justice, are involved at almost all other levels of governance, including the urban level, where the Commission seeks to stimulate policy activity. Furthermore, the EU's relationship to national environmental policy retains significant hierarchical elements. The European Union determines criteria and permits variation in implementation but it will still enforce a failure to transpose EU directives. It is worth noting that weakening of national authority does not necessarily imply a reduction of hierarchy, if developments in governance simply mean the European Union replaces national governments at the top of the pyramid.

Although areas of EU governance, particularly water, are being transformed, this is not a spontaneous, bottom-up process. The 2000 EU Water Framework Directive is breaking new ground in managing water on a river basin basis and involving stakeholders in creating basin management plans. But paradoxically, the European Union is *compelling*

Member States to govern in a more decentralized fashion. Member States do not have the option to *not* submit river basin management plans. Thus while we observe an increase in policy making activity on the river basin level, it demonstrates the strength of the authority of the European Union, not a diminution of it. There would be evidence of weakening of hierarchy and the EU's authority if river basin management plans were voluntary, if Member States routinely failed to create river basin management plans and if failures to comply went unpunished.

Of the cases examined here, Canada comes to closest to fitting the decentring and fragmentation that MLG Theory describes. Even in areas of exclusive federal jurisdiction, the Canadian federal government consults extensively with provincial governments, such as the negotiation of trade agreements. The federal government is very sparing in using its authority vis-à-vis provincial governments, often forgoing using powers it has constitutionally (Watts 2006, 328). In environmental policy particularly, Canadian governments negotiate as partners, in a network, without hierarchy. Writing about hierarchical decision making, Scharpf (1997, 173) observed that '... the reach of hierarchical authority is constrained by the exit options that are available to the target audience.' Because provinces can refuse to enter into any federal provincial agreement, they retain the option to exit. This is not the case with most environmental policies in the EU, US, Australia or Germany.

In water allocation policy, the Canadian federal government sees its role as a facilitator of agreements, not an enforcer of those agreements. (Despite this, as noted above, as of 2013 only one of Canada's interprovincial rivers is covered by an interprovincial agreement, although negotiations are ongoing for at least one other river.) With regard to climate policy, Canada approximates the MLG theory ideal, with extensive horizontal linkages. Barry Rabe (2007) has described the exhaustive process of stakeholder consultations the Canadian federal government has engaged in over climate policy, in addition to years of federal-provincial negotiations, both bilateral and multilateral. In environmental policy, the Canadian federal government is emphatically *not* a superior level of government, one with authority over provincial governments (Weibust 2009).

WEAKENING OF STATE AUTHORITY IN RELATION TO NON-STATE ACTORS

With regard to the claims of MLG theory concerning the reduced importance of state authority to the benefit of non-state actors, Heinmiller's chapter is the most relevant. While non-state actors play some role in influencing policy in the other cases, it is only in the irrigation policy cases that non-state actors are key, playing a direct role in developments. This characterization does not strictly fit the MLG framework, however. First these non-state actors play the role they do because of irrigation rights that were created by law and granted to them. These groups do not fit a pluralist model of organic representation of interests from civil society at large.

Furthermore, in discussing horizontal weakening of state relations, MLG theory implies some fluidity and unpredictability in the interplay of actors. Heinmiller, by contrast, argues that governance and participation in irrigation policy are far from 'fluid.' These represent iron triangles of policy formulation, where particular private groups have a stable, long standing and exclusive relationship with government authorities. This formulation of the relationship is also at odds with MLG theory's claim that the horizontal weakening of authority is beneficial. Since the term was coined decades ago, iron triangles have been considered a pathological form of governance, whereby public resources are monopolized for the benefit of select private interests (Lowi 1969).

BEYOND MLG: VARIETIES OF MLG AND THE SHADOW OF HIERARCHY

The cases presented here represent several distinct variants of multilevel governance, only some of which fit the MLG theory typology. Some cases fit with existing federal typologies, but others do not. If the MLG theory doesn't fit what we observe, can we turn to older federal models?

With regard to water policy several cases here fit the *administrative federalism* model: the higher level of government makes policy which the lower level has discretion in how to implement. The lower level of government does not however, have the choice *not* to implement. There is no possibility of subnational veto or exit. The lower level of government has flexibility, but not autonomy. This model adequately represents much of water quality policy in the US and the EU. Canadian water and irrigation policy largely fit the *layer cake* (or dual) model of federalism of two distinct and wholly separate levels of government. Each level has areas of exclusive jurisdiction and operates almost wholly autonomously within those areas of jurisdiction.

Unfortunately, existing federal typologies of governance are inadequate for many of the other cases, particularly cases of climate policy. First, most federal typologies pertain to only two levels of government and exclude the local level of government. Second, they cannot adequately incorporate the international level, which can be important for water governance and is of paramount importance for climate change. Third, even where there are just two relevant levels of government, existing models do not reflect the variety of patterns of interaction between the levels, which can include elements of hierarchy as well as cooperation. For example, in American climate policy international developments play a role, as do cooperative initiatives between American states, in addition to state and federal policies.

Even if federal models are amplified by the addition of local governments as a 'junior' level of government to the subnational (or EU Member State level), they do not adequately describe the relationships. A greater shortcoming is the inability of traditional federal accounts to incorporate the international level, which is particularly significant for climate change. International cooperation whether it be multilateral (such as the UN Framework Convention on Climate Change process), plurilateral (US, EU and China) or bilateral (US and China, US and Canada) will significantly condition policy choices at the federal or EU level. As Douglas Macdonald observed in Chapter 9, the Canadian federal government's preoccupation with harmonizing with any US federal climate policy and the EU's desire for a global leadership role illustrate the importance of incorporating the international into models of national or EU policy making.

Absent another suitable term to describe these complex governance arrangements, it seems likely that the term 'multilevel governance' will continue in use. As we have seen, the term is already employed in contexts far removed from the European Union and entirely divorced from Hooghe and Marks' more elaborate theoretical apparatus. 'Multilevel governments' is not a satisfactory substitute because, while the international and supranational levels clearly constitute levels of governance, they are not governments. The vagueness and ambiguity of the term remains a virtue, as it can encompass a wide variety of real world governance arrangements.

PERFORMANCE OF MULTILEVEL GOVERNANCE

If a weak central state is beneficial on normative grounds, then these benefits should be more evident in Canada than other MLG systems. The Canadian federal system is among the world's most decentralized (Watts 2006, 327), particularly in environmental governance (Weibust 2009). Furthermore, the Canadian federal system probably comes closer to any other in approximating network governance, rather than hierarchy.

In the cases of climate and water policy presented here, the Canadian federal state is relatively weak or almost entirely absent. In water policy, there is no evidence of the exercise of federal power by the Canadian government. In climate policy, the federal government's extensive consultation with provincial governments and commitments that no region should bear a disproportionate impact of climate policy, left very little room to manoeuvre. The exception here would be if the federal government leapfrogged the process of federal-provincial negotiation in climate policy and simply emulated US policy. Macdonald argues in Chapter 9 that this is what Prime Minister Stephen Harper sought to do. That would represent a weakening of national authority (and a very unpopular ceding of Canadian sovereignty). But, unlike the ideal of MLG, it would not represent network government or a dispersion of authority. Rather it would represent an inversion of Robert Putnam's *Two-Level Games* (1988), whereby to escape the trap of decision making by executive federalism, the federal government accedes to another country's legislation, rendering the decision immune to provincial re-negotiation.

The cases presented here do not support the claim that making governance less hierarchical, more decentralized, is beneficial for environmental outcomes. If negotiation driven network-based governance produced superior environmental protection, Canadian environmental policy would be the most effective. On the contrary, the cases presented here lend weight to previous findings that hierarchy, while not sufficient for effective environmental governance, can be very helpful to it. As economic theories of optimal jurisdiction predict, fragmented governance is not well suited to addressing externalities that cross borders. Thus the presence of a higher level of government (or a higher court) is conducive to the resolution of water allocation disputes and ensuring that some water is reserved for instream ecological benefits, rather than allocating as much as possible for consumptive uses, such as irrigation. This pattern is also consistent with the findings of Layzer, Beierle and Cayford (described in the Introduction) that a backstop of hierarchy was more advantageous for sustainable ecosystem management in the United States than wholly autonomous state or local level decision making, even though many of those cases did not exhibit the substantial transboundary spillovers seen in climate change.

The argument here is not a call for centralization, for all policy to be made and implemented by the federal or EU level. In air and water pollution control the prevailing American and European Union arrangement of devolved implementation of minimum standards, coupled with central oversight, appears to have been very effective (Lowry 1992).

Decisions do not need to be imposed hierarchically, but the existence of the shadow of hierarchy permits the benefits of negotiation and decentralization, while mitigating problems such as externalities and conflict resolution (Scharpf 1997).

NOTE

1. American federal authority intrudes through jurisdiction over key dams, the Endangered Species Act, and rights of Indian Tribes, to name a few.

REFERENCES

Beierle, Thomas C. and Jerry Cayford (2002), *Democracy in Practice: Public Participation in Environmental Decisions*, Washington DC: Resources for the Future.

Börzel, Tanja (2010), 'European Governance: Negotiation and Competition in the Shadow Of Hierarchy,' *Journal of Common Market Studies* **48**(2), 191–219.

Braun, D. (2006), 'Between Market Preserving Federalism and Intergovernmental Coordination: The Case of Australia,' *Swiss Political Science Review*, **12**(2), 1–36

Ellerman, D and B. Buchner (2008), 'Over-Allocation or Abatement? A Preliminary Analysis of the EU ETS Based on the 2005-06 Emissions Data,' *Environmental and Resource Economics* **41**(2), 267–87.

European Commission (2013), 'Structural reform of the European carbon market,' available at http://ec.europa.eu/clima/policies/ets/reform/index_en.htm (accessed 24 June 2013).

Harrison, Kathryn (2007), 'The Road not Taken: Climate Change Policy in Canada and the United States,' *Global Environmental Politics* **7**(4), 92–117.

Harrison, Kathryn and Lisa Sundstrom (2007), 'The Comparative Politics of Climate Change,' *Global Environmental Politics* **7**(4), 1–18.

Heyes, Anthony (2000), 'Implementing Environmental Regulation: Enforcement and Compliance,' *Journal of Regulatory Economics* **17**(2), 107–29.

Hooghe, Liesbet and Gary Marks (2001), *Multi-Level Governance and European Integration*, Lanham: Rowman and Littlefield.

Koontz, Tomas M. (2002), *Federalism in the Forest: National versus State Natural Resource Policy*, Washington DC: Georgetown University Press.

Layzer, Judith A. (2008), *Natural Experiments: Ecosystem-Based Management and the Environment*, Cambridge: The MIT Press.

Lowi, Theodore (1969), *The End of Liberalism,* New York: Norton.

Peters, B. Guy (2006), 'Federalism and Public Administration: the United States and the European Union,' in Anand Menon and Martin Schain (eds) *Comparative Federalism: the European Union and the United States in Comparative Perspective*, Oxford: Oxford University Press, pp. 177–94.

Putnam, Robert (1988), 'Diplomacy and Domestic Politics: the Logic of Two-Level Games,' *International Organization* 42(3), 427–60.

Rabe, Barry G. (2007), 'Beyond Kyoto: Climate Change Policy in Multilevel Governance Systems,' *Governance* **20**(3), 423–44.
Rabe, Barry G. (2009), 'Second-Generation Climate Policies in the States: Proliferation, Diffusion and Regionalization,' in Henrik Selin and Stacy D. VanDeveer (eds) *Changing Climates in North American Politics: Institutions, Policymaking and Multilevel Governance*, Cambridge: MIT Press, pp. 67–86.
Scharpf, Fritz (1997), *Games Real Actors Play: Actor-Centered Institutionalism in Policy Research*, Boulder: Westview Press.
Schneider, Hans-Peter (2006), 'The Federal Republic of Germany,' in Akhtar Majeed, Ronald L. Watts and Douglas M. Brown (eds) *Distribution of Powers and Responsibilities in Federal Countries*, Montreal and Kingston: McGill-Queen's University Press, pp. 124–54.
Simpson, Jeffrey, Mark Jaccard and Nic Rivers (2007), *Hot Air: Meeting Canada's Climate Change Challenge*, Toronto: McClelland and Stewart.
Watts, Ronald L. (2006), 'Comparative Conclusions,' in Akhtar Majeed, Ronald L. Watts and Douglas M. Brown (eds) *Distribution of Powers and Responsibilities in Federal Countries*, Montreal and Kingston: McGill-Queen's University Press, pp. 322–49.
Weibust, Inger (2009), *Green Leviathan: The Case for a Federal Role in Environmental Policy*, Burlington: Ashgate.

Index

Althusius, Johannes
theory of federalism 41
Australia 9, 11, 16, 40, 58, 61, 155–60, 174–5, 182–3, 259, 262, 264
Australian Capital Territory 68
Carbon Pollution Reduction Scheme (CPRS) 163–5
COAG Water Reform Framework (1994) 68
coal mining industry in 182
Commonwealth Water Act 70
Constitution of 63
Council of Australian Governments (COAG) 159, 161–2, 164–5, 169–75
Council of the Australian Federation (CAF) 161, 163, 166, 171–3
emissions trading scheme (ETS) 162–3
federal government of 59
Murray-Darling Basin 63, 67, 69
Murray-Darling Basin Initiative (1990) 68
Murray-Darling Basin Plan 70
National Emissions Trading Scheme (NETS) 163, 171, 172
National Emissions Trading Taskforce (NETT) 162
National Plan for Water Security 70
National Water Commission 70
New South Wales 11, 59–60, 63–4, 68–9, 71, 75, 164
Queensland 68, 164
River, Bow 60
River, Murrumbidgee 60
River, Snowy 69
River, St Mary 60, 62
River Murray Waters Agreement (1914) 63
Victoria 68, 164
working group on climate change and water (WGCCW) 163–4, 170–71, 174
Austria 191

Belgium 88, 93, 191
Brussels 6, 90, 111, 120–22
failure to report RBMP 89
Berlin Rules (2004)
shortcomings of 92
Brandeis, Louis
concept of 'laboratories of democracy' 7
Bulgaria 88, 90, 190
accession to EU (2007) 91
Burden Sharing Agreement
provisions of 188
Bush, George H.W.
environmental policies of 74
Bush, George W.
administration of 145
environmental policies of 136, 140
opposition to Kyoto Protocol 136

Calvinism 41
Cameron, David 118
Canada 7, 9, 11–12, 16, 20, 58, 61, 81, 95, 101, 103–4, 154–60, 174–5, 183, 186, 193, 198–9, 256–7, 260, 266–7
Alberta 11, 59, 62–5, 67, 71, 75, 95–100, 159, 166–9, 199, 201–3, 205–6, 256, 258–60, 262
Albert Water Act (1999) 66
British Columbia 96, 137, 150, 169, 200–201, 203, 263
Calgary 96

Canadian Council of Ministers of the Environment (CCME) 166
Canadian Environmental Protection Act 168
constitution of 201
Constitution Act (1867) 62
Council of Energy Ministers (CEM) 166
Council of the Federation (COF) 166
energy infrastructure of 202
Environment Canada 167
First Ministers Meeting (FMM) 165–6, 168, 170, 173
government of 65
Joint Ministers Meeting (JMM) 166–7, 169, 171–4
Lethbridge 96
Mackenzie River Basin Board (MRBB) 97, 102
Manitoba 137, 169
Medicine Hat 96
National Action Program on Climate Change (1995) 167, 199
National Air Issues Coordinating Committee – Climate Change (NAICC-CC) 166
National Climate Change Process (NCCP) 166–7, 169, 171–2, 174
National Climate Change Secretariat (NCCS) 166
National Implementation Strategy and Business Plan 199
Natural Resources Canada 167
Northwest Territories 96–7
Nova Scotia 202
oil and gas industry of 182
Ontario 137, 168–9, 200–201, 203
Ottawa 62
Peace/Athabasca/Slave Watershed 97
Prairie Provinces Advisory Water Board 98
Prairie Provinces Water Board (PPWB) 96–9, 101–2
Progressive Conservatives 65
Quebec 137, 150, 159, 167–9, 199, 201, 203, 257, 263
Red Deer 96, 99
River, Mackenzie 94, 103–4
River, Oldman 65
River, Peace 94
River, North Saskatchewan 94, 101
River, South Saskatchewan (SSR) 94–6, 99, 101, 103, 260
Saskatchewan 62, 96, 98–9, 202–3, 205, 256
Saskatoon 96
Water Resources Act (1930) 65
Western Water Board Agreement (WWBA) 98
Yukon Territory 96
Carbon Capture and Storage (CCS) 172
Carter, Jimmy
environmental policies of 61
Catholicism
social theory of 42
China 132, 266
Chrétien, Jean 170, 204
Canadian Prime Minister 167
Climate Alliance 111
aims for GHG reduction 253–4
Conference of Parties (CoP) 131
Berlin (1995) 193
Copenhagen (2009) 187, 189, 200
Kyoto (1997) 187–8
corporatism 20
Cyprus 88

deforestation 18
Delors, Jacques
European Commission President 252
Denmark 188, 191
Copenhagen 125, 127

Energy Cities 124
offices of 122
Environmental Kuznets Curve (EKC)
concept of 17
research 17–18
environmentalism 71
political influence of 72–3
European Coal and Steel Community (ECSC)
founding of (1951) 42
European Commission (EC) 12, 46–7, 111–12, 204–5, 222, 224, 227–8, 230–31, 234–5, 237–43, 250–51

Annual Survey 240
Bathing Water Directive 48–50, 52
Covenant of Mayors (2008) 111, 121–3, 125–6
Dangerous Substances in Water Directive 48
Drinking Water Directive 49
Environmental Action Programme (EAP) 43–4, 47–8
Fifth Environmental Action Programme (1993) 229–30, 233–4
Floods Directive (2007) 50
Fourth Environmental Action Programme (1987) 49, 227–30
funding provided by 215
Laeken Council Declaration (2001) 46
Large Combustion Plant Directive (1988) 44
Marine Strategy Directive (2008) 50
Nitrates Directive 52
personnel of 252
Sixth Environmental Action Programme (6EAP) 235–6, 241
Urban Wasterwater Treatment Directive 53
White Paper on European Governance (2001) 46, 120
European Community Treaty 232
provisions of 12, 222–3, 237
European Council 189, 191, 229, 232, 235, 237, 257
Dublin Declaration (1990) 228
Edinburgh Meeting (1992) 46
European Court of Justice 12, 51, 93, 223–4, 233, 239, 241–2, 250, 263
Community law 216–21, 225–6, 229, 232
European Economic Community (EEC)
Council of Minsters 43
Paris Council (1972) 43
Treaty of Rome (1957) 42–3
European Environment Agency 51, 89
European Network for the Implementation and Enforcement of Environmental Law (IMPEL) 236, 241
European Parliament 50, 190, 227, 229, 236–8
Committee on Legal Affairs 241
European Union (EU) 3–4, 6–7, 9–12, 16, 29, 39–41, 47–8, 51, 54, 81, 87, 103, 111–13, 120–21, 182–6, 198–9, 206–7, 215, 239, 243, 250, 252, 256, 258, 260, 264–7
Aarhus Convention 236
as example of MLG 4
Biofuels Directive 117
Co-generation Directive 117
Convention on Long-Range Transboundary Air Pollution (LRTAP) 26
Emissions Trading Scheme (ETS) 148, 189–90, 194–5, 204, 258, 262–3
Great Lakes Regional Collaboration 26
Lisbon Treaty (2009) 47, 191
Maastricht Treaty (1992) 44–5, 225
member states of 4–5, 12, 22, 40, 43, 45, 47, 49, 51–3, 87–91, 93–4, 113–18, 121, 123, 184, 187–93, 195, 201, 205, 215–16, 218–23, 225–8, 230–33, 235–40, 242–3, 250–52, 257, 261–4
Renewable Electricity Directive 117
Single European Act (SEA) (1986) 40, 44, 116, 190
Treaty of Amsterdam (ToA) (1997) 46, 190
Treaty on European Union (TEU) (1993) 190, 216–17
Water Framework Directive (WFD) 50, 88–93, 104, 252, 261, 263–4

Federal Republic of Germany (West Germany)
environmental policies of 44
federalism 16, 19–20, 24, 41, 76, 144, 155
administrative 265
American 141
Canadian 25, 155, 159, 202
collaborative 147–8, 255–6
executive 165

German 196
Federation of German Industry 194
Finland 191
'First in Time, First in Right' (FITFIR) 83, 96, 98, 259–60
concept of 82
First World War (1914–18) 123
Ford, Gerald
administration of 135
France 89, 191

de Gaulle, Charles
temporary withdrawal of France from EEC 43
German Democratic Republic (East Germany) 193
economy of 194
Germany 9, 12, 16, 21, 88, 118, 120, 123, 183, 186, 191, 193–5, 198–9, 206, 258, 264
Baden-Wuttemberg 195
Bavaria 195
Brandenburg 197
Bundesrat 184, 196–7, 204, 257
coal industries of 182
Conference of the Environmental Ministers (1973) 197
First National Climate Change Mitigation Program 194
GHG emissions of 192, 194, 197
Hessen 195
Lander 192, 194–8, 201, 204, 253, 257
Leitfaden Kommunaler Klimaschutz 119
Lower Saxony 195
Munich 125, 127
North-Rhine Westphalia (NRW) 119–20, 194–5, 197–8, 204–5
Renewable Energy Act (REA) (2000) 194–5, 199
reunification of 193
Saarland 197
Gillard, Julia 164
environmental policies of 70, 164
Global Financial Crisis (2007–9) 258
Granholm, Jennifer
Governor of Michigan 131
Greece 90–91, 188, 191
failure to report RBMP 89
greenhouse gas (GHG) emissions 18, 117–18, 149, 151, 158, 184, 257, 261–2
reduction costs 12, 181, 187, 202–4
reduction targets for 10, 12, 111, 116, 119–20, 122, 125, 127, 131, 136–9, 141, 146, 181, 189, 191, 200, 202, 253, 256
vehicular 145

Harper, Stephen
administration of 200, 203
hierarchical climate governance 112–17, 120–21
shortcomings of 125–6
horizontal climate governance 112, 115–16, 123, 125, 162
effects of 112
project networks 124
sister city arrangements 124
transnational city networks 124, 126
Howard, John
administration of 69–70
environmental policies of 163

intergovernmental relations (IGR) 12, 60, 121, 123, 133, 154–5, 159, 165, 168–9, 175, 183, 185, 187
centralization of 117
horizontal 162, 166
institutions of 154–6, 161, 173–4, 184
intergovernmental agreement (IGA) 161, 172
Integrated Water Resources Management (IWRM) 100, 103
promotion of 92
variants of 87
Intelligent Energy Europe (IEE) 122
aims of 121
International Freshwater Treaties Database 90
International Joint Commission (IJC)
participants in 62
Italy 88, 189, 191
Rome 242

Japan 21

Kohl, Helmut
  administration of 193
Koontz, Tomas M.
  Federalism in the Forest (2002) 64
Kyoto Protocol (1997) 117, 134–5, 167, 170, 200
  provisions of 181, 253–4, 257–8
  ratification of (2002) 199–200, 203–4
  signatories of 136, 162, 168

Labor Party (Australia)
  members of 163–4
land degradation 18
Layzer, J.A.
  Natural Experiments (2008) 64
Lithuania
  GHG emissions of 192
Luxembourg 191

Malta 88
Martin, Paul
  administration of 168
micro-regional governance
  examples of 26
Merkel, Angela
  administration of 198
Mesta/Nestos Agreement 90–91, 94
  shortcomings of 93
Meuse/Scheldt Agreement 94
Mulroney, Brian
  Canadian Prime Minister 166–7
multilevel governance (MLG) 4, 10, 15–16, 20, 31, 102, 112–13, 116, 155, 249–50, 253–4, 262–5, 267
  concept of 4, 6, 264
  impact on environmental performance 22–3
  lack of focus on horizontal climate governance 115
  patterns of 16–17, 27

National Round Table on Environment and Economy
  report of (2012) 200
neo-corporatism 21–2
Netherlands 93, 123, 191
  Amsterdam 125, 127
  Antwerp 90
  Bestuursakkord Nieuwe Stijl (BANS) Agreement 119, 125
  Klimaatcovenant 118–19
  Rotterdam 90
  Stimulering LOkale Klimaatinitiatieven (SLOK) program 119
Nichols, Mary
  California Air Resources Board Chair 138
non-governmental organisations (NGOs) 113
North, Douglass
  concept of 'adaptive efficiency' 151

Obama, Barack 151, 168
  administration of 140–42, 145
  environmental policies of 131, 141, 254

Poland 189–90
  economy of 192
  GHG emissions of 192
  Warsaw 242
policy cycle stages 24–5
  agenda setting 23–4
  change 23
  evaluation 23, 25
  formulation 23–4
  implementation 23–5
  legitimation 23
pollution 17, 21, 27–8, 44, 51–3, 60, 220
  air 17, 166
  control policies 18–19, 25, 163–4
  water 17, 48–9, 61, 252
Portugal 188, 191
  failure to report RBMP 89
  Lisbon 242

Reagan, Ronald
  environmental policies of 25, 135
Republic of Ireland 191
Rio Summit 116–17

River Basin Management Plans (RBMP) 89
international 91
River, Danube
RBMP for 91
River, Elbe
RBMP for 91
River, Ems
RBMP for 91
River, Meuse 90–91
RBMP for 91
River, Nestos 90–91
River, Oder
RBMP for 91
River, Rhine
RBMP for 91
River, Scheldt
RBMP for 91
River, Shannon
RBMP for 91
Romania 190
GHG emissions of 192
Rudd, Kevin
electoral victory of (2007) 163
environmental policies of 70
Russian Federation
natural gas reserves of 193

Schroeder, Gerhard
administration of 194
Schwarzenegger, Arnold 145
Governor of California 131, 139
Second World War (1939–45) 123–4
shadow of hierarchy
concept of 8
examples of 255, 260, 265, 268
Slovenia
GHG emissions of 192
Social Democrat Party (Germany) 198
Spain 9, 88, 123, 191
Madrid 242
River, Ebro 89, 93
water irrigation in 88
subnational governance
concept of 16
subsidiarity 40–42, 53
concept of 40
use as scaling device 52
Sweden 21, 190–91
Stockholm 125, 127

Thatcher, Margaret
economic policies of 45
Triptych Approach 189, 193
concept of 188

United Kingdom (UK) 19, 21, 88–9, 117, 188, 191, 258
British North America Act (1867) 200–201
CO_2 reduction in 117–18
government of 45, 48
Parliament 200
United Nations (UN)
ECE Convention on the Protection and Use of Transboundary Watercourses and International Lakes 91
Framework Convention on Climate Change (UNFCCC) (1992) 131, 167, 187, 193, 199, 203, 266
United States of America (US) 7, 9, 11, 16, 19–22, 40, 58, 61, 81–2, 103–4, 133–4, 136, 147, 160, 182–3, 186, 250, 260, 264–8
Alabama 81, 84
American Clean Energy and Security Act (ACESA) 146
Apalachicola-Chattahoochee-Flint (ACF) river system 84–6
Arizona 80, 83
CALFED initiative 73–4
California 9–11, 24, 26–7, 59–60, 64, 67, 72–3, 75–6, 80, 135, 137, 143, 145, 149–50, 152, 259–60, 262
Central Valley Project Improvement Act (1992) 74
Chesapeake Bay Commission 26–7
Clean Air Act 28, 140–41, 145, 255
Clean Water Act 28, 72
Climate Security Act 145
coal mining industry in 182
Congress 20, 74, 81–5, 136–7, 141, 143–4, 146, 149, 151, 159, 184
Constitution of 64, 136

Endangered Species Act 72–4, 82, 85
Environmental Council of the States 142
Environmental Protection Agency (EPA) 140, 148–9
Environmental Water Account 73, 76
federal government of 59
Florida 81, 84
Georgia 81, 84
GHG emissions of 131–2
Global Climate Protection Act (1987) 135
Global Warming Solutions Act (2006) 24–5
Harmon Doctrine 81, 101
Metropolitan Water District 72
Midwestern Greenhouse Gas Reduction Accord (MGGRA) 137
National Ambient Air Quality Standards 28
National Association of Clean Air Agencies 142, 144
National Association of State Energy Officials 142
National Conference of State Legislatures (NCSL) 142
National Environmental Performance Partnership System 148
National Governors Association 142
Northeast States for Coordinated Air Use Management 26
Pecos River Compact (1948) 83
Regional Greenhouse Gas Initiative (RGGI) 25–6, 29, 137, 140, 148, 150
renewable portfolio standards (RPS) 136, 147, 152
Republican Party 135, 143
River, Colorado 80, 82–3
Safe Drinking Water Act Primary Standards 28
State and Local Climate Change Program 255
South Florida Water Management District 26
State Water Resources Control Board 60
Supreme Court 7, 80, 82–5, 132, 140, 145, 260–61
US Army Corp of Engineers 72
Washington DC 72
Western Climate Initiative (WCI) 137, 149, 200

vertical climate governance 112, 114–15, 118, 125–6
concept of 120–22
variants of 123

water allocation 59–62, 89–90
constitutional policies regarding 63–4
interstate 80–82, 85–6
large-scale 69, 75
regional policies 65–7
water scarcity 9–10, 63, 68, 87, 95, 150
aims to prevent 89, 94
Wilson, Pete
Governor of California 74